MECHANICAL ENGINEERING THEORY AND APPLICATIONS SERIES

MECHANICAL VIBRATIONS OF DISCONTINUOUS SYSTEMS

MECHANICAL ENGINEERING THEORY AND APPLICATIONS SERIES

Mechanical Vibrations of Discontinuous Systems
Juan Carlos Jauregui-Correa and Ocar M. Gonzalez Brambila
2009. ISBN: 978-1-60876-126-5

MECHANICAL ENGINEERING THEORY AND APPLICATIONS SERIES

MECHANICAL VIBRATIONS OF DISCONTINUOUS SYSTEMS

JUAN CARLOS JAUREGUI-CORREA

AND

OSCAR M. GONZALEZ-BRAMBILA

Nova Science Publishers, Inc.

New York

For permission to use material from this book please contact us:
Telephone 631-231-7269; Fax 631-231-8175
Web Site: http://www.novapublishers.com

LIBRARY OF CONGRESS CATALOGING-IN-PUBLICATION DATA

ISBN 978-1-60876-126-5

Available upon request

Published by Nova Science Publishers, Inc. ✦ New York

CONTENTS

Introduction		**vii**
Chapter 1	Nonlinear Vibrations	**1**
Chapter 2	Mechanical Systems with Stiffness Discontinuities	**33**
Chapter 3	Stiffness Discontinuities due to Large Deformations	**63**
Chapter 4	Mechanical Systems that Combine Soft and Hard Materials	**93**
Chapter 5	Solution Methods	**111**
References		**133**
Index		**137**

INTRODUCTION

Mechanical vibrations have attracted the attention of engineers and researchers a long time ago. Vibration problems require a deep understanding of mathematical modeling and signal analysis, but they also require practical experience. Even though there is abundant literature about the subject, there are still many unknown answers from modeling through measurement interpretation. In this book, we want to contribute to these plentiful sources of information. The scope of it is the identification of discontinuities in the stiffness properties of mechanical systems, and how these discontinuities affect its behavior. We want to find, with a practical perspective, those parameters that cause nonlinearities, or modify the stiffness properties.

We define a nonlinear system when one of its elements is modeled as a nonlinear function. We will show that some systems that present a discontinuity in their properties can be treated as nonlinear functions. Others can be modified in such a way that the discontinuity can be modeled as part of the damping properties.

We have identified that the discontinuities of a mechanical system can be associated to:

- System's displacement
- System's velocity
- Material properties

Stiffness and damping nonlinearities depend upon temperature, frequency and excitation amplitudes. They can have stiffness functions of the form kx^3 , or damping functions as $(av^2+bx^2)sing(v)$. If we considered friction damping, normally we use Coulomb's equation that will have the form of: μN if $v > 0$ and $-\mu N$ if $v < 0$.

Piecewise linear dynamical systems are used to analyze vibrations in elements such as gears, rolling bearings, sliding bearings, rotor bearing, elastic-plastic and viscoelastic structures. In these cases, stiffness variations are a function of the system's displacements, and they are discontinuous. It means that there is an abrupt change in the stiffness function as the system moves. These cases can be described with Duffing's equation. Although most discontinuities in a system's parameters cause nonlinearities, there are some nonlinear systems that have no discontinuities in their properties. In most cases, systems with discontinuities show jump phenomenon behavior, and some times they can behave chaotically.

For those systems that have discontinuities associated to system's velocity, their behavior will depend upon the stiffness coefficients. In mechanical systems, they usually depend upon the excitation frequency rather than the system's output velocity. Therefore, velocity dependant terms are directly related to damping, frequency dependant terms are related to stiffness nonlinearities.

Material discontinuities make systems to present different behaviors. Very slender elements have large deformations which produces stiffness nonlinearities that behave as a Duffing's type equation. Other systems have smooth stiffness changes; although they have nonlinear coefficients, their output behavior shows a time delay response that can be treated as an artificial damping. The other type of material stiffness discontinuities are found in mechanical systems that combine soft and hard materials. Most of the passive vibration isolators are built combining soft materials (normally a rubber type material) with metals. For this type of stiffness discontinuities, we present an approach based on the viscoelasticity theory. In this way, we found an alternative method for modeling such systems with the aid of the Finite Element Method.

As we can see, some sort of nonlinear behavior is found in every mechanical system, but in most cases they can be approximated as linear systems achieving sufficiently good solutions. Although it is the fastest way to solving a design problem, it produces erroneous solutions when nonlinear behaviors are present. Nonlinear problems are a lot more difficult to solve, and not always we can find a solution.

We have to improve the way we model nonlinear mechanical systems, because modern machinery operates at higher speeds and they are design with lighter elements. These two aspects of modern machines respond to the need of having more efficient systems, but, on the other hand, they show nonlinearities at frequencies closer to the operating velocities, making the analysis and measurement's interpretation harder. Another characteristic of modern machinery is that their nonlinearities exhibit a chaotic behavior. This means that the output response depends upon initial conditions. Thus, with small variations at the input, we obtained a totally different solution. In many ways, this behavior is treated as random solution

Any mechanical system can be represented as a set of differential equation; these equations can be derived using different approaches, such as a Newtonian modeling, or they can be found from the Hamilton's principle. These differential equations can be classified as linear or nonlinear. The difference between them is the nature of their terms. It means that, if the coefficients are constant, the differential equation is linear, if they are a function of one of the variables, they are nonlinear. In this book, we described the analysis and dynamic response of mechanical systems that can be represented by nonlinear differential equations with the characteristic that the stiffness has a discontinuous function.

There is a wide range of mechanical nonlinear systems that can be represented as a lumped mass system, connected through springs and dampers, where the nonlinear terms are associated with the passive elements. The nonlinear characteristic is represented by approximating either the springs or the dampers as polynomial functions instead of simple constants. Thus, nonlinear stiffness or dampers are simulated mathematically in such a way that its effects can be understood and extrapolated into practical cases.

In chapter 1 we introduce the concept of nonlinear vibrations, we describe different methods for modeling nonlinear systems and we include the most common analysis methods. In chapter 2 we present models for mechanical systems that show stiffness discontinuities. We also show how models vary among different elements such as: gears, bearings, sliding bearings, or rotor rubbing. For such systems, we describe how they can be modeled as a lump-mass system, and how to incorporate the discontinuities into the stiffness model.

Continuous systems also show stiffness discontinuities; thus, chapter 3 describes systems that present stiffness variations due to large deformations. These variations create a model with a similar pattern as a Duffing's equation; therefore, they can be treated as systems with stiffness discontinuities. Other continuous systems that show stiffness discontinuities are those that combine soft and hard materials. Chapter 4 includes these type of systems, and we describe its application to passive isolators.

In chapter 5 we present some of the most useful analysis methods for nonlinear systems. We include: system identification, phase diagram analysis, and time-frequency maps. These methods are a lot more useful than the traditional spectral analysis, because they consider more information than the spectral analysis. Also, these methods will be included into modern vibration monitoring systems in a near future.

It is important to say that basic vibration theory and fundamentals of the Finite Element Method are out of the scope of this book. We recommend the reader to review those topics before.

Chapter 1

NONLINEAR VIBRATIONS

BACKGROUND

Dynamics started after Newton's laws and the solution of differential equations; the simplicity of his equations and the ability to describe complex problems has made them very useful until today. Newton's idea of a universe moving as a perfect machine leaded to the idea of a predictable universe in time and space. At the end of the XIX Century, Poincaré found that Newton's law could not predict the trajectory of three bodies (Sun, Moon and Earth) since small differences in their initial conditions resulted in totally different solutions. He found that, although there was no a final solution for the position of each body, the system was stable within a solution space well limited.

Until the evolution of computers, it was possible to understand Poincaré's, Hadamard's and Duhem's discoveries and formulations. Edward Lorenz, in 1963, developed a mathematical model that tried to predict the weather, his results showed a very sophisticated behavior of three relatively simple differential equations. Even more, their solution differ from each other with slightly small initial conditions; thus, even small errors from field measurements resulted in a different solution. Lorenz called this sensitivity to initial conditions as "divergences from close trajectories"; but latter on it was named "chaos".

René Thorn found that a chaotic system shows unstable equilibrium after which it bifurcates into a new physical behavior, then the new configuration could develop new bifurcations and so on. Thus, bifurcations are the most important characteristic of a chaotic system. An example of bifurcation is the simple buckling problem of an axially loaded beam; it could bend indistinctly toward one or the other side (Figure 1.1).

STATE SPACE DYNAMICS

In physics, a "state space" is a complex Hilbert space within which the possible instantaneous states of the system may be described by a unit vector. It was originally introduced by Poincaré at the end of the XIX century, and it is at this time when the concept of limit cycle was introduced. To explain this concept, we present its application to a well-known system: simple harmonic oscillator. This system is autonomous since it has no explicit time, t, dependency.

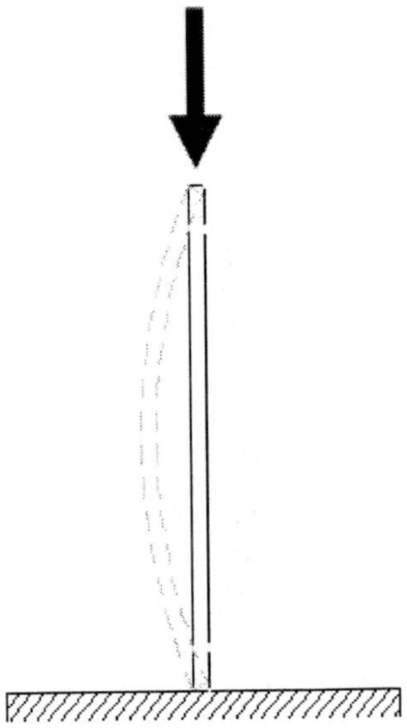

Figure 1.1. Example of a bifurcation system: A slender bar under compression load, the bar will buckle to either left or right side.

Its differential equation can be stated as:

$$\ddot{x} + \omega_0^2 x = 0 \tag{1.1}$$

This has a solution of the form of

$$\begin{aligned} x &= A\cos\omega_0\, t + B\sin\omega_0\, t \\ &= x_0\cos\omega_0\, t + \frac{x_0}{\omega_0}\sin\omega_0\, t \end{aligned} \tag{1.2}$$

This solution can be expressed as

$$\begin{aligned} x &= C\cos(\omega_0 t + \phi) \\ \dot{x} &= -C\omega_0\sin(\omega_0 t + \phi) \end{aligned} \tag{1.3}$$

Where C and ϕ depend upon initial conditions:

$$C = \sqrt{A^2 + B^2} = \sqrt{x_0^2 + \frac{\dot{x}_0^2}{\omega_0^2}}$$

$$\tan\phi = -\frac{B}{A} = -\frac{\dot{x}_0}{\omega_0 x_0} \tag{1.4}$$

The phase diagram is a graphical representation of the function $\dot{x} = f(x)$. In this way, the system's response is represented independently of time as $S = S(x, \dot{x})$. The trajectory of any point in this function is known as phase trajectory. Since C, ω_o and ϕ are constants and depend upon the system parameters and the initial conditions; thus, the shape of the phase diagram is an ellipse of the form (Figure 1.2)

$$x^2 + \left(\frac{\dot{x}}{\omega_0}\right)^2 = C^2 \tag{1.5}$$

Another representation of the phase diagram is a graphical plot of the function $S = S(x, \dot{x}, t)$

There is another concept related to the phase diagram: radius vector. This vector defines the phase rate of change, and it is expressed as:

$$\mathbf{v} = \frac{d}{dt}\mathbf{r} = \mathbf{i}\dot{x} + \mathbf{j}\ddot{x}$$
$$= \mathbf{i}\{-C\,\omega_0 \sin(\omega_0 t + \phi)\} + \mathbf{j}\{-C\,\omega_0^2 \cos(\omega_0 t + \phi)\} \tag{1.6}$$

It is clear that the phase rate is always positive and for a linear system it is monotonic. In Chapter 5 we will present different forms of the phase diagram. In this representation, the linear system produces closed orbits every period $T = 2\pi/\omega_o$. A phase diagram has the advantage that it can be found without integrating the differential equation. In this case, the diagram can be found from the state variable concept. If we substitute

$x_1 = x$ and $x_2 = \dot{x}$, then:

$$\dot{x}_1 = x_2$$
$$\dot{x}_2 = -\omega_0^2\, x \tag{1.7}$$

Thus, dividing these two equations, a first order differential equation is found.

$$\frac{d\,x_2}{d\,x_1} = -\omega_0^2\,\frac{x_1}{x_2} \tag{1.8}$$

The phase diagram has an elliptic shape with a slope of the form

$$\frac{d\dot{x}}{dx} = -\frac{\omega_0 x}{\sqrt{(x_0)^2 - x^2}}$$

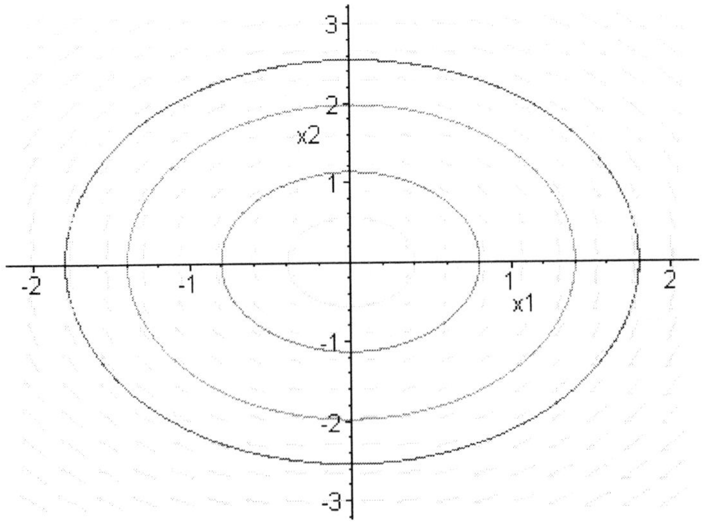

Figure 1.2. Phase diagram for the family solution of the simple harmonic oscillator $\ddot{x} + \omega_0^2 x = 0$.

From this equation, it can be seen that there is limited solution space bounded by the initial condition x_0.

Higher order systems show a wider dynamic behavior, thus the phase diagram trajectories have larger solution margins. Thus, a nonlinear oscillator of the form $\ddot{x} + f(x,\dot{x}) = 0$, can be represented as a system of two first order equations:

$$\begin{aligned} \dot{x}_1 &= f_1(x_1, x_2) \\ \dot{x}_2 &= f_2(x_1, x_2) \end{aligned} \tag{1.9}$$

Where f_1 and f_2 are nonlinear functions. The stability of the system can be determined through the analysis of the phase diagram trajectories of the function $S = S(x,\dot{x})$, and the fixed point X_1 and X_2 are those that satisfies:

$$\begin{aligned} f_1(X_1, X_2) &= 0 \\ f_2(X_1, X_2) &= 0 \end{aligned} \tag{1.10}$$

In most cases, the solution of equation 1.10 is only possible with numerical technique such as Newton-Raphson's methods. Now, if the solution starts at a relatively closed fixed point, the stability analysis is similar to a linear system. If we define a new variable *y*, such that:

$$y_1 = x_1 - X_1$$
$$y_2 = x_2 - X_2 \tag{1.11}$$

which can be understand as a small perturbation around the fixed point X. With the aid of this perturbation, it is possible to find the stability of the system. It will depend on the propagation of the perturbation. Thus, the dynamic system will have the form of:

$$\begin{aligned} \dot{y}_1 &= \dot{x}_1 = f_1(X_1 + y_1, X_2 + y_2) \\ \dot{y}_2 &= \dot{x}_2 = f_2(X_1 + y_1, X_2 + y_2) \end{aligned} \tag{1.12}$$

Expressing the equation as a Taylor's series:

$$\begin{aligned} \dot{y}_1 &= f_1(X_1, X_2) + y_1 \frac{\partial f_1}{\partial x_1} + y_2 \frac{\partial f_1}{\partial x_2} + O(y_1^2, y_2^2, y_1 y_2) \\ &= y_1 \frac{\partial f_1}{\partial x_1} + y_2 \frac{\partial f_1}{\partial x_2} + O(y_1^2, y_2^2, y_1 y_2) \end{aligned} \tag{1.13}$$

and

$$\dot{y}_2 = y_1 \frac{\partial f_2}{\partial x_1} + y_2 \frac{\partial f_2}{\partial x_2} + O(y_1^2, y_2^2, y_1 y_2) \tag{1.14}$$

Eliminating higher order terms, the perturbed system can be expressed as a linear system of the form:

$$\{y\} = [A]\{y\} \tag{1.15}$$

Where $[A]$ is the Jacobean matrix around X_1 and X_2

$$\mathbf{A} = \begin{bmatrix} \dfrac{\partial f_1}{\partial x_1} & \dfrac{\partial f_1}{\partial x_2} \\ \\ \dfrac{\partial f_2}{\partial x_1} & \dfrac{\partial f_2}{\partial x_2} \end{bmatrix}_{(X_1, X_2)} \tag{1.16}$$

The eigenvalues of matrix $[A]$ define the system's stability.

$$\lambda_{1,2} = \frac{1}{2}\left(\tau \pm \sqrt{\tau^2 - 4\Delta}\right) \tag{1.17}$$

Where Δ is the determinant of matrix $[A]$, and τ is its trace.
The number of fixed points can be attractors, rejecters or they can oscillate.

1. Stable point: They asymptotically attract any solution. They exist if $\Delta > 0$, $\tau < 0$, and $\tau^2 - 4\Delta > 0$.
2. Instable point: They asymptotically reject any solution. They are present when $\Delta > 0$, $\tau > 0$, and $\tau^2 - 4\Delta > 0$.
3. Saddle point: They attract any solution from one side, and reject them from the other. The only condition is $\Delta < 0$.
4. Stable spiral: The attract any solution and keep oscillating, they are defined when $\Delta > 0$, $\tau < 0$, and $\tau^2 - 4\Delta < 0$.
5. Instable spiral: The reject any solution in and oscillating manner, they are present when $\Delta > 0$, $\tau > 0$, y $\tau^2 - 4\Delta < 0$.
6. Limit cycle: They represent those equilibrium points around which every trajectory remains oscillating around it. There is no attraction or rejection and their condition is that $\Delta > 0$
7. Liapunov nodes: They are a solution where $\Delta = 0$, therefore, at least one eigenvalue is cero. Thus, there is no a single fixed point, but a series of points.
8. Star nodes: They are those particular points for the case when $\tau^2 - 4\Delta = 0$. In this case, it is necessary to have a particular analysis.

We will discuss some of the most common systems. To begin with, we will analyze the linear oscillator define by the dynamic equation: $\ddot{x} + \omega_0^2 x = 0$. Then, if we transform this equation in a state system:

$$f_1(x_1, x_2) = x_2$$
$$f_2(x_1, x_2) = -\omega_0^2 x_1$$

The Jacobean matrix for this system is

$$[A] = \begin{bmatrix} 0 & 1 \\ -\omega_0^2 & 0 \end{bmatrix}$$

Following the analysis we can see that there is a fixed point at $X_1 = 0$ and $X_2 = 0$. It is also important to notice that $\Delta = \omega_0^2$ and $\tau = 0$. Therefore, we can conclude that the system has a limit cycle at $X_1 = 0$ and $X_2 = 0$, and its phase diagram is a constant ellipse with center at this point.

Another example is the van der Pol equation. This equation has the following expression:

$$\ddot{x} + \varepsilon\left(x^2 - 1\right)\dot{x} + \omega_0^2\, x = 0 \tag{1.18}$$

The state system will have the form of:

$$\begin{aligned} f_1(x_1, x_2) &= x_2 \\ f_2(x_1, x_2) &= -\varepsilon(x_1^2 - 1)x_2 - \omega_0^2 x_1 \end{aligned} \tag{1.19}$$

And the Jacobean matrix:

$$[A] = \begin{bmatrix} 0 & 1 \\ -2\varepsilon x_1 x_2 - \omega_0^2 & -\varepsilon x_1^2 + \varepsilon \end{bmatrix} \tag{1.20}$$

According to equation 1.10, there is only one fixed point at $X_1=0$ and $X_2=0$. Thus, the trace is $\tau = \varepsilon$, and its determinant is $\Delta = \omega_0^2$. For the van der Pol equation, the phase diagram will show a close loop, but it won't be an ellipse and the time response will be periodic but not sinusoidal. These important characteristics will be discussed in Chapter 5.

Most mechanical systems we include in this book can be represented with Duffing's nonlinear oscillator. In general, a Duffing's oscillator has the form:

$$\ddot{x} + \omega_0^2 x + \alpha x^3 = 0 \tag{1.21}$$

Its state functions will be:

$$\begin{aligned} f_1(x_1, x_2) &= x_2 \\ f_2(x_1, x_2) &= -\omega_0^2 x_1 - \alpha x_1^3 \end{aligned} \tag{1.22}$$

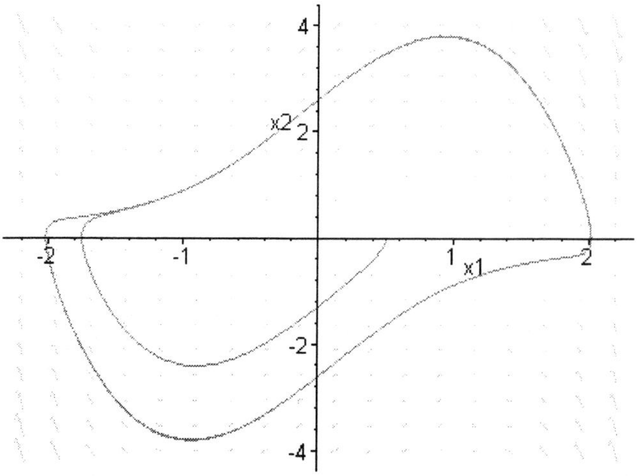

Figure 1.3. Phase diagram for the van der Pol equation.

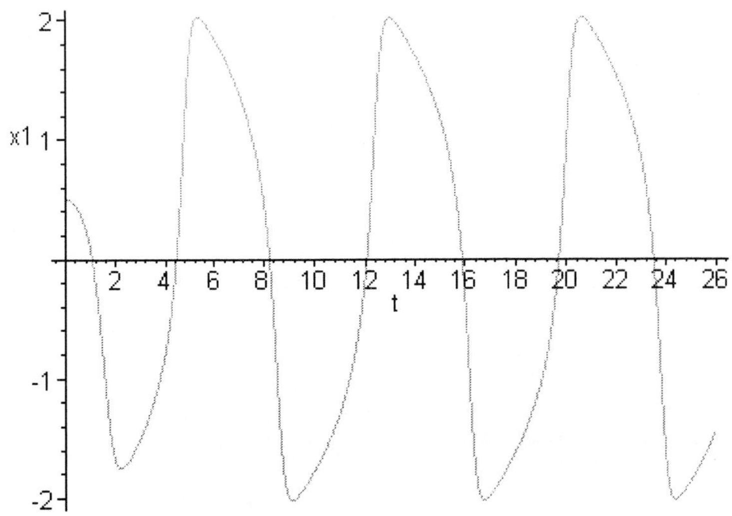

Figure 1.4. Time solution for the van der Pol equation.

And the Jacobean matrix:

$$[A] = \begin{bmatrix} 0 & 1 \\ -3\alpha x_1^2 - \omega_0^2 & 0 \end{bmatrix} \quad (1.23)$$

In this case there is no single fixed point. Then, from equation 1.10, the fixed points are:

$$(X_1, X_2) = \begin{Bmatrix} (0,0) \\ \left(\dfrac{\omega_0}{\sqrt{-\alpha}}, 0\right) \\ \left(\dfrac{-\omega_0}{\sqrt{-\alpha}}, 0\right) \end{Bmatrix} \quad (1.24)$$

There are four possible cases, which will depend upon α's sign. If α is positive, the system shows a softening stiffness, if α is negative, the system has a hardening stiffness.

In order to illustrate the analysis, we present an example where $\omega = 1$, $\alpha = 0.25$ and the initial conditions are $x(0) = 1, \dot{x}(0) = 0$.

In this case, the fixed points are located at:

$$(X_1, X_2) = \{(0,0), (2i,0), (-2i,0)\} \quad (1.25)$$

The phase diagram is represented in Figure 1.5, and the displacement as a function of time in Figure 1.6.

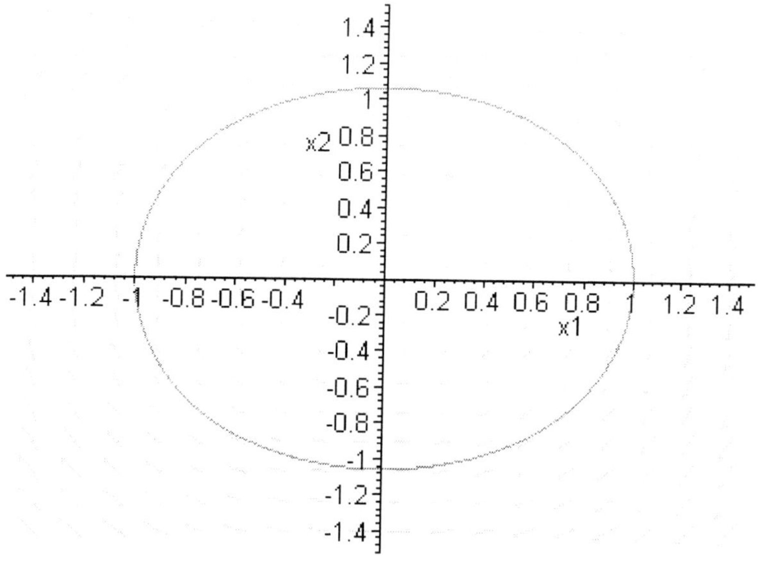

Figure 1.5. Phase diagram for the system $\ddot{x} + x + 0.25\,x^3 = 0$

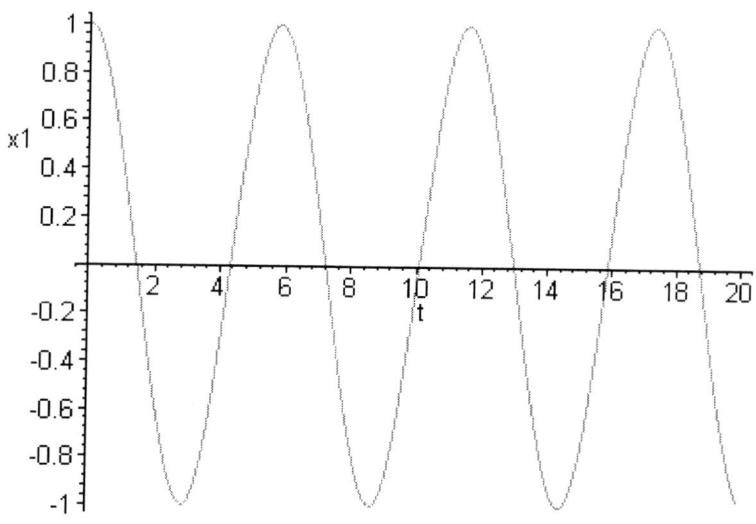

Figure 1.6. Time solution for the system $\ddot{x} + x + 0.25\,x^3 = 0$

It has a similar response as a linear oscillator, since the eigenvalues of the Jacobean matrix are the same at point *(0,0)*. It is clear α has a minimum contribution to the system. In order to see the nonlinear effects, we analyze a system with $\alpha = -1$, in which case the fixed points are located at:

$$(X_1, X_2) = \{(0,0), (\omega_0, 0), (-\omega_0, 0)\} \tag{1.26}$$

The system will have stable solutions when the initial condition starts within the range $-\omega_o < x_0 < \omega_o$ and the parameter $\omega_o > 0$. Any other point outside this vicinity will show instable trajectories. It is important to point out that even if the phase diagram has a closed form, it is not an ellipse and the system's response is periodic but not sinusoidal. This important characteristic will be use in Chapter 5, where the harmonic distortion concept will be explained. The phase diagram is shown in Figure 1.7 and the solution as a function of time in Figure 1.18.

BIFURCATIONS

Fixed points have a particular situation; they depend upon system's parameters. This implies that vector's qualitative structure changes as parameters vary with time. In this way, fixed point are created, destroyed, or its stability modified. The name of these qualitative changes are called local bifurcations, since they can be analyzed locally near the fixed point for a first order system, or near the limit cycle for a second order system.

A good clear example of bifurcation is column buckling. A thin column can sustain a compression load beyond certain critical limit, the column will keep its straight shape unless a small perturbation modifies its configuration; at this instant the column will suddenly buckle. The bifurcation comes from the fact that the column could buckle to any direction along the perpendicular plane of the applied force. In order to understand bifurcations, the Multiple-Center Theorem establishes that the effective dimension number of bifurcations depends upon the characteristic exponents that are equal to zero. In contrast, in a stable system, the real parts are negative; and in an unstable system, they are positive.

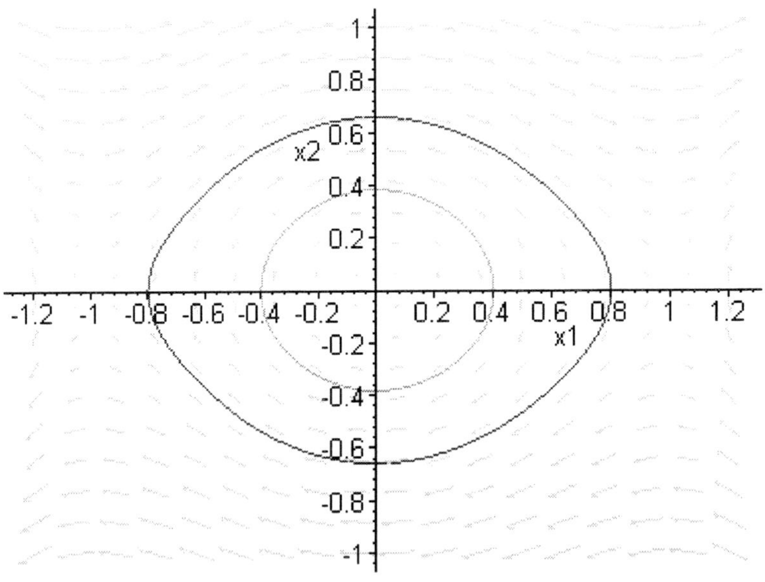

Figure 1.7. Phase diagram for system $\ddot{x} + x - x^3 = 0$

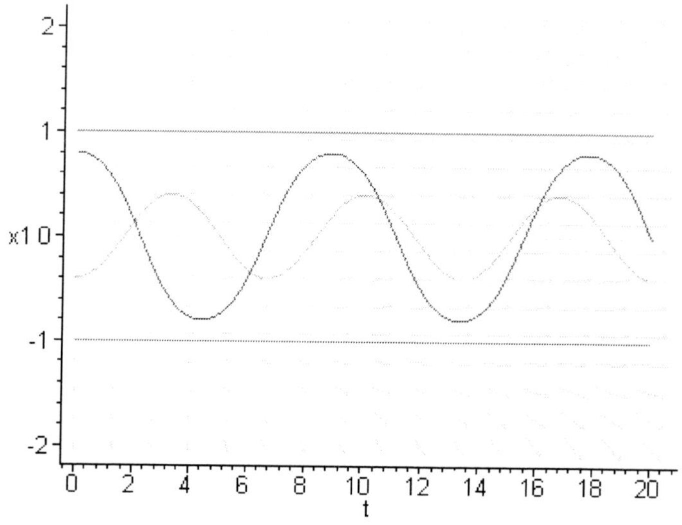

Figure 1.8. Time solution for system $\ddot{x} + x - x^3 = 0$.

A simple way of their classification is through the reduction of the differential equation to its normal form:

$$\frac{dx}{dt} = A_0 + B_0 x + C_0 x^2 + \dots$$
$$+ r\left(A_1 + B_1 x + C_1 x^2 + \dots\right)$$
$$+ r^2\left(A_2 + B_2 x + C_2 x^2 + \dots\right)$$
$$+ \dots \tag{1.27}$$

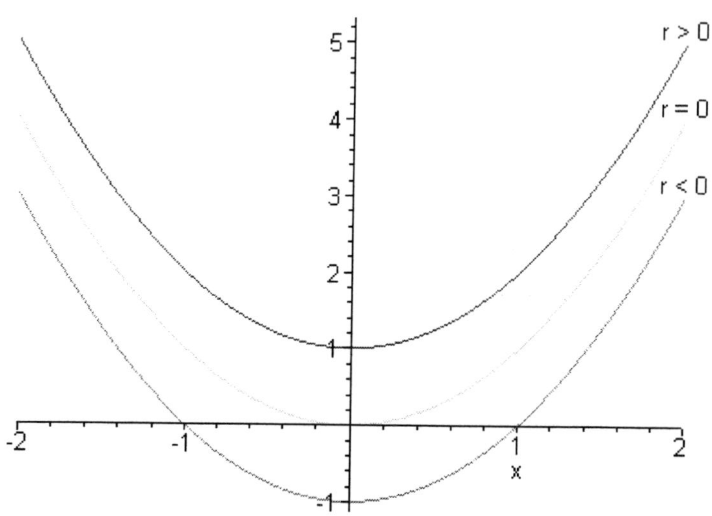

Figure 1.9. Sketch of a saddle bifurcation.

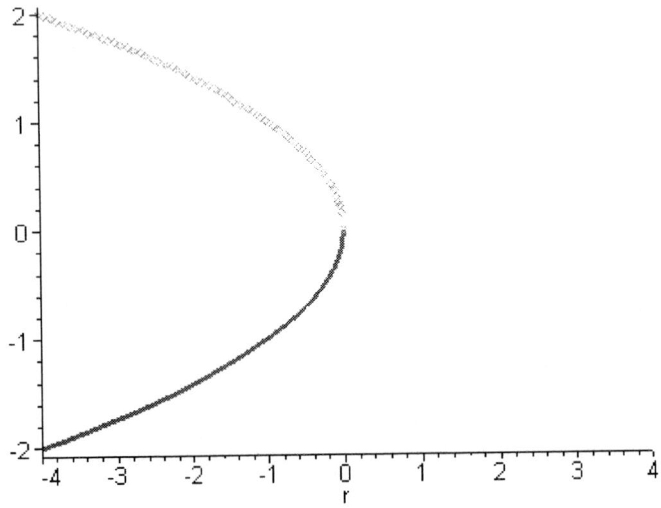

Figure 1.10. Bifurcation diagram of a Saddle bifurcation.

Where *A*, *B* and *C* define the magnitude of the control parameter *r*. With the parameter's values, it is possible to classify bifurcations.

Saddle Bifurcation

The prototype of this bifurcation is a first order system such as:

$$\dot{x} = r + x^2 \tag{1.28}$$

When the system parameters are modified, the fixed points coincide and collapse. When *r* is negative, there are two fixed points, one is stable and the other is unstable. As *r* approaches zero, the fixed points get together and form a half-stable point that vanishes when *r* changes to a positive value. In this case, bifurcation appears at $r = 0$, because vectors for $r < 0$ and $r > 0$ are completely different.

Trans-Critical Bifurcation

There are some systems, which have a fixed point at any parameter, and it cannot be destroyed; but at certain parameter variations, the fixed point's stability or structure changes as a function these variations. A typical differential equation for this type of system is

$$\dot{x} = r x - x^2 \tag{1.29}$$

In this case, there is a fixed point at $X=0$; nevertheless, its stability changes because for $r < 0$ the fixed point is stable, whereas for $r > 0$ it is unstable. It can be noticed that there are two fixed points regardless of r $(r = 0)$.

Super Critical Bifurcation

This bifurcation appears in those systems that are symmetrical with respect to the origin, such as a column under axial compression. A typical differential equation for these systems will be:

$$\dot{x} = r\,x - x^3 \tag{1.30}$$

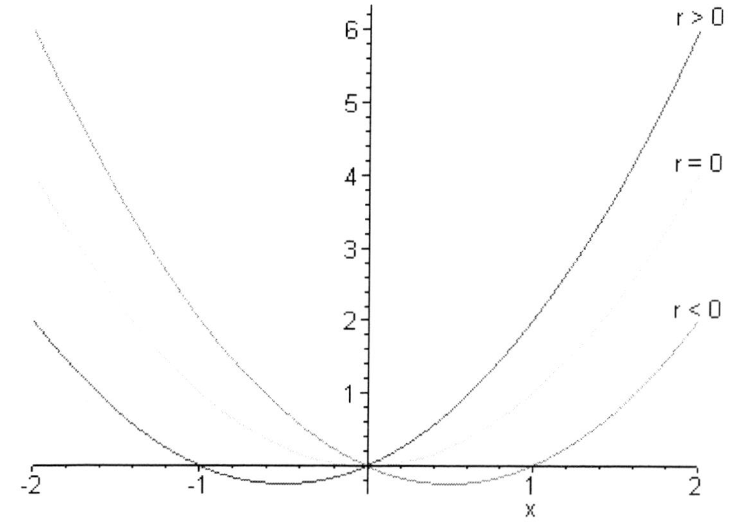

Figure 1.11. Sketch of a trans-critical bifurcation.

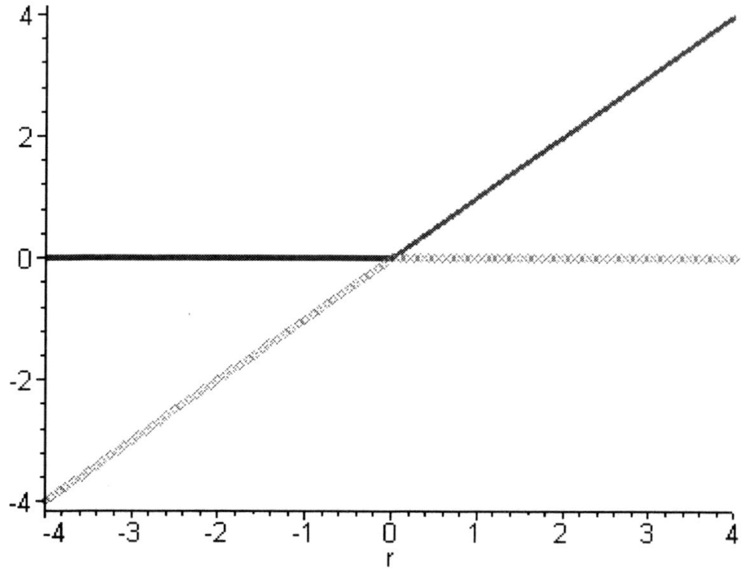

Figure 1.12. Bifurcation diagram of a trans-critical bifurcation.

This equation is invariant, which means that $f(-x) = -f(x)$. If $r < 0$, the cubic term acts as a restoring force and there is only one fixed point that is stable. When $r = 0$ the origin is stable, but the system shows a logarithmic decrement. Finally, when $r > 0$, the origin is unstable and there two new fixed points at $X = \pm\sqrt{r}$.

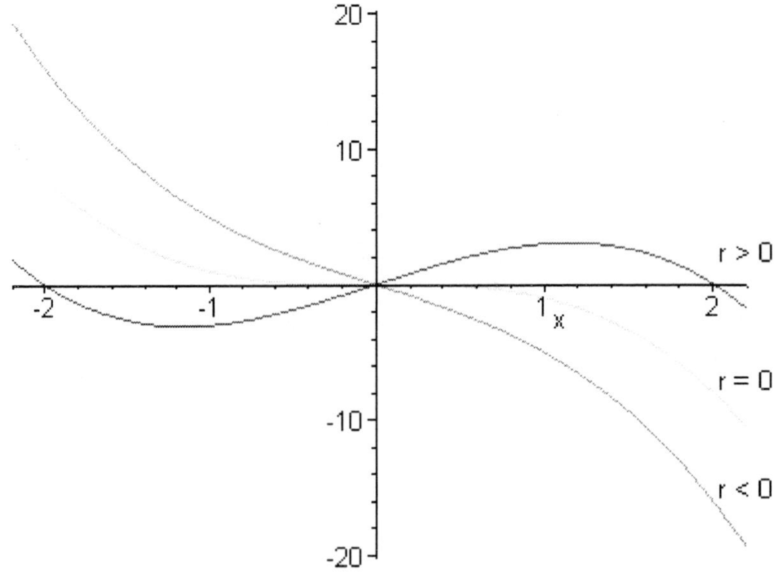

Figure 1.13. Sketch of a super critical bifurcation.

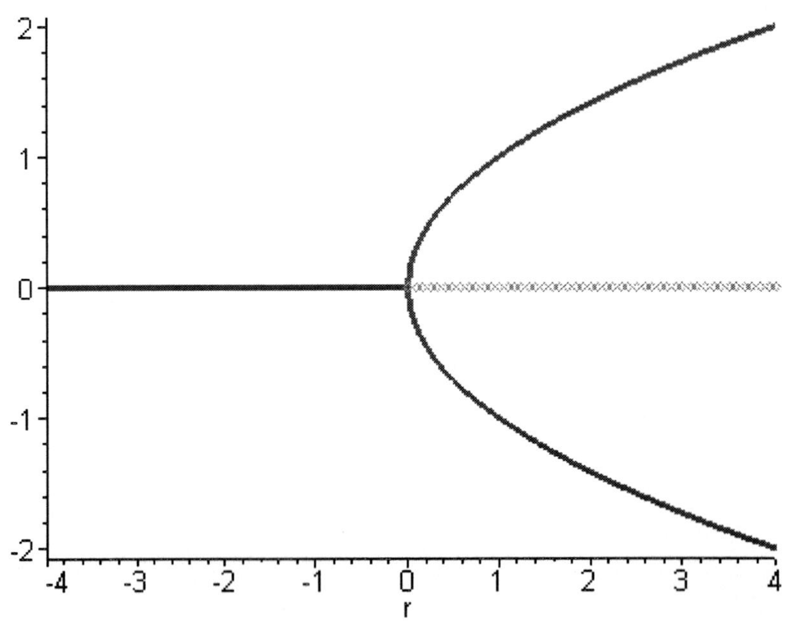

Figure 1.14. Bifurcation diagram of a super critical bifurcation.

Sub Critical Bifurcation

Whereas in a super critical bifurcation the cubic term stabilizes the system, in this case it makes it unstable. A typical differential equation will have the form of:

$$\dot{x} = r\,x + x^3 \tag{1.31}$$

The origin is stable for $r < 0$ and unstable for $r > 0$; but since there is no restoring term, trajectories tend to infinite at any initial condition $x_o \neq 0$.

Since there is no a physical representation of a system with such equation, it should have a restoring force of the form:

$$\dot{x} = r\,x + x^3 - x^5 \tag{1.32}$$

Two-Scale Method

Strogatz presented this method for the first time in 1994; it is a modification to the regular perturbation theory, which show effectiveness when it is applied to first order differential equations but fails with second order systems. Its application to oscillating system with small non-linearity shows good results, and it is done if the original system is modified as:

$$\ddot{x} + \omega_0^2 x + \varepsilon\, h(x, \dot{x}) = 0 \tag{1.33}$$

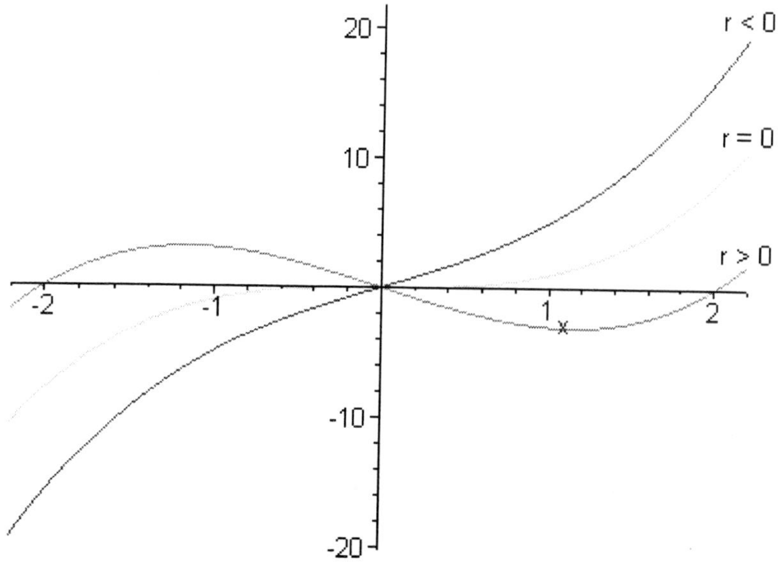

Figure 1.15. Sketch of a sub critical bifurcation.

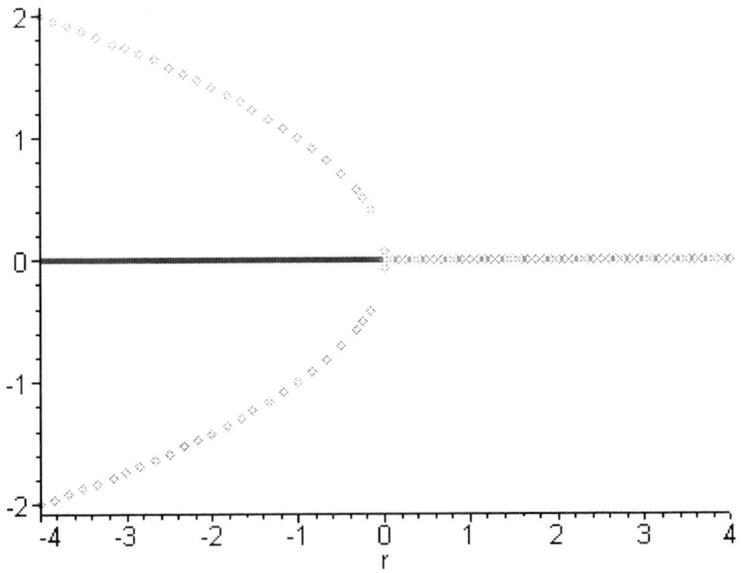

Figure 1.16. Bifurcation diagram of a sub critical bifurcation.

Where ε is a small positive number and $h(x,\dot{x})$ is an arbitrary smooth function.

The Two-Scale Method assumes that the non-linear response varies at a slower rate compared to the fundamental cycle; therefore, including two-time scales into the system's response, the solution can be expanded as:

$$x(t,\varepsilon) = x_0(\tau,T) + \varepsilon\, x_1(\tau,T) + O(\varepsilon^2) \tag{1.34}$$

The slow response is $T=\varepsilon t$, and the fast one is $\tau = t$. The first derivative with respect to time is expressed as:

$$\dot{x} = \frac{dx}{dt} = \frac{\partial x}{\partial \tau} + \frac{\partial x}{\partial T}\frac{\partial T}{\partial t} = \frac{\partial x}{\partial \tau} + \varepsilon \frac{\partial x}{\partial T} \tag{1.35}$$

Substituting equation 1.34 into 1.35, the first derivate becomes:

$$\begin{aligned}\dot{x} &= \frac{\partial}{\partial \tau}(x_0 + \varepsilon x_1) + \varepsilon \frac{\partial}{\partial T}(x_0 + \varepsilon x_1) + O(\varepsilon^2) \\ &= \frac{\partial x_0}{\partial \tau} + \varepsilon\left(\frac{\partial x_1}{\partial \tau} + \frac{\partial x_0}{\partial T}\right) + O(\varepsilon^2)\end{aligned} \tag{1.36}$$

Similarly, the second derivative is:

$$\ddot{x} = \frac{d\dot{x}}{dt} = \frac{\partial \dot{x}}{\partial \tau} + \varepsilon \frac{\partial \dot{x}}{\partial T} + O(\varepsilon^2)$$
$$= \frac{\partial^2 x_0}{\partial \tau^2} + \varepsilon \left(\frac{\partial^2 x_1}{\partial \tau^2} + 2\frac{\partial^2 x_0}{\partial \tau \partial T} \right) + O(\varepsilon^2) \tag{1.37}$$

Finally, the system becomes:

$$\frac{\partial^2 x_0}{\partial \tau^2} + \varepsilon \left(\frac{\partial^2 x_1}{\partial \tau^2} + 2\frac{\partial^2 x_1}{\partial \tau \partial T} \right) + \omega^2 x_0 + \varepsilon h\left(x_0, \frac{\partial x_0}{\partial \tau} \right) = 0 \tag{1.38}$$

Rearranging terms around x_1 and x_0:

$$\frac{\partial^2 x_0}{\partial \tau^2} + \omega^2 x_0 = 0 \tag{1.39}$$

$$\frac{\partial^2 x_1}{\partial \tau^2} + \omega^2 x_1 = -2\frac{\partial^2 x_0}{\partial \tau \partial T} - h\left(x_0, \frac{\partial x_0}{\partial \tau} \right) \tag{1.40}$$

Equation 1.39 is a simple oscillator with a solution of the form:

$$x_0 = r(T)\cos(\omega_0 \tau + \phi(T)) \tag{1.41}$$

Where $r(T)$ and $\phi(T)$ are function of the slow term. For the fast term τ they are kept constant. If we substitute them into equation (1.40), the second term becomes:

$$\frac{\partial^2 x_1}{\partial \tau^2} + \omega^2 x_1 = 2\omega_0 (r'\sin(\omega_0 \tau + \phi(t)) + r\phi'\cos(\omega_0 \tau + \phi(t))) - h \tag{1.42}$$

Where h is a function of the form

$$h = h(r(\tau)\cos(\omega_0 \tau + \phi(T)), -r\omega_0 \sin(\omega_0 \tau + \phi(T))) \tag{1.43}$$

And, if we substitute $\theta = \omega_0 \tau + \phi$, and approximating h by a Fourier's series, $h(\theta)$ can be obtained from the Fourier coefficients with the following identity:

$$\frac{1}{2\pi} \int_0^{2\pi} h(\theta) f(\theta) d\theta \equiv \langle h f(\theta) \rangle \tag{1.44}$$

where:

$$h(\theta) = \sum_{k=0}^{\infty} a_k \cos(k\theta) + \sum_{k=1}^{\infty} b_k \sin(k\theta)$$

Then, equation 1.42 becomes

$$\partial_{\tau\tau} x_1 + \omega_0^2 x_1 = 2\omega_0 \left[r' \sin(\theta) + r\phi' \cos(\theta) \right] \\ - \sum_{k=0}^{\infty} a_k \cos(k\theta) - \sum_{k=1}^{\infty} b_k \sin(k\theta) \tag{1.45}$$

Rearranging terms, we found that for $k = 1$:

$$(2r'\omega_0 - b_1)\sin\theta$$
$$(2r\phi'\omega_0 - a_1)\cos\theta$$

And, in order to avoid resonance, they have to satisfy

$$(2r'\omega_0 - b_1) = 0$$
$$(2r\phi'\omega_0 - a_1) = 0$$

In this way, r' and $r\phi'$ are determined from:

$$r' = \frac{b_1}{2\omega_0} = \frac{1}{2\pi\omega_0} \int_0^{2\pi} h(\theta) \sin\theta \, d\theta \equiv \left\langle \frac{h}{\omega} \sin\theta \right\rangle \tag{1.46}$$

And

$$r\phi' = \frac{a_1}{2\omega_0} = \frac{1}{2\pi\omega_0} \int_0^{2\pi} h(\theta) \cos\theta \, d\theta \equiv \left\langle \frac{h}{\omega} \cos\theta \right\rangle$$

The most common integrals $\langle h(\theta) \rangle$ are listed in the appendix at the end of the chapter.

They are the average solution for every cycle of θ. In other words, they are the "average" or the "slow time" solution of the fast variable θ, and the slow variable r is constant. We can get the solution on real time t, when we solve r' and $r\phi'$, and then we substitute them in equation 1.41.

If we apply this method for an approximate solution $x(t, \varepsilon)$ to a damped simple oscillator with the following initial conditions:

$$\ddot{x} + 2\varepsilon\dot{x} + \omega_0^2 x = 0$$
$$x(0) = 1, \dot{x}(0) = 0$$

We can see that the damped simple oscillator satisfies the conditions for an autonomous system. From equation 1.41

$$x = r(T)\cos(\omega_0\,\tau + \phi(T))$$

where h has the form of

$$h = 2\dot{x} = -2\omega_0\, r\sin(\omega_0\,\tau + \phi)$$

and substituting h into equation 1.46, we find that:

$$r' = \frac{1}{\omega}\langle h\sin(\omega_0\,\tau + \phi)\rangle$$
$$= \frac{1}{\omega}\langle -2\omega_0\, r\sin(\omega_0\,\tau + \phi)\sin(\omega_0\,\tau + \phi)\rangle$$
$$= -2r\underbrace{\langle \sin^2(\omega_0\,\tau + \phi)\rangle}_{=1/2}$$
$$= -r$$

In the same way, substituting h into equation 1.47 we find that:

$$r\dot{\phi} = \frac{1}{\omega_0}\langle h\cos(\omega_0\,\tau + \phi)\rangle$$
$$= \frac{1}{\omega}\langle -2\omega_0\, r\sin(\omega_0\,\tau + \phi)\cos(\omega_0\,\tau + \phi)\rangle$$
$$= 2r\underbrace{\langle \sin(\omega_0\,\tau + \phi)\cos(\omega_0\,\tau + \phi)\rangle}_{=0}$$
$$= 0$$

If we apply the variable separation method for the solution of r, and we integrate for ϕ, we find that

$$r = c_1\,\mathbf{e}^{-T}$$
$$\phi = c_2$$

c_1 and c_2 are determined when we apply the initial conditions.

$$r(0) \approx \sqrt{x(0)^2 + \dot{x}(0)^2} = 1 \Rightarrow c_1 = 1$$

$$\phi(0) \approx \tan^{-1}\left(\frac{\dot{x}(0)}{x(0)}\right) = 0 \Rightarrow c_2 = 0$$

Since we defined that $\tau = t$, and $T = \varepsilon t$, the final solution is:

$$\begin{aligned} x &= \mathbf{e}^{-T} \cos(\omega_0 \tau) \\ &= \mathbf{e}^{-\varepsilon t} \cos(\omega_0 t) \end{aligned}$$

Figures 1.17 and 1.18 show the solution for a case where $\varepsilon = 0.5$. Even if ε is relativity large, the approximation of the Two-Scale Method gives very good results, if we compare them with the exact solution.

The application of the Two-Scale Method to a damped simple oscillator helps us to estimate the deviation from a known solution. Since the results are acceptable, we apply an approximate solution $x(t, \varepsilon)$ to a nonlinear oscillator with the following initial conditions:

$$\ddot{x} + x + \varepsilon x^3 = 0$$

$$x(0) = 1, \dot{x}(0) = 0$$

Similarly as in previous case, we define $h = x^3$; then, equation 1.41 becomes:

$$h = r^3 \cos^3 \theta$$

and equations 1.46 and 1.47 are:

$$r' = \left\langle \frac{h}{\omega} \sin\theta \right\rangle = \frac{r^3}{\omega} \underbrace{\left\langle \cos^3\theta \sin\theta \right\rangle}_{=0} = 0$$

$$r\phi' = \left\langle \frac{h}{\omega} \cos\theta \right\rangle = \frac{r^3}{\omega} \underbrace{\left\langle \cos^4\theta \right\rangle}_{=3/8} = \frac{3}{8} r^3$$

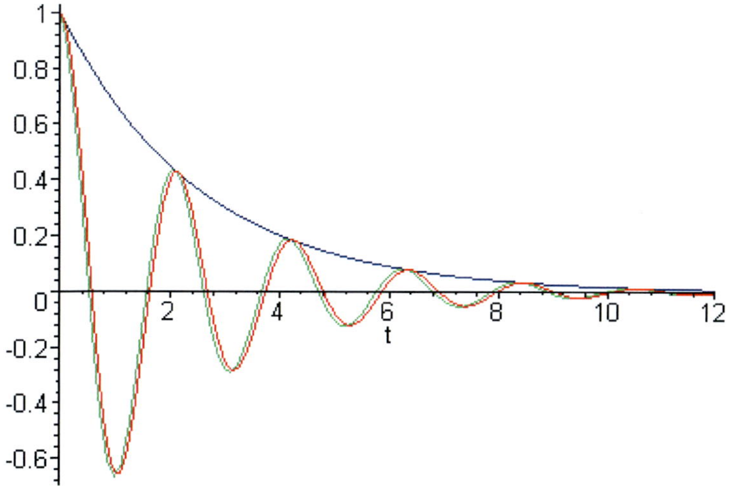

Figure 1.17. Solution for the simple damped oscillator with $\varepsilon = 0.5$. The blue line corresponds to r, the green line is the approximate solution, and the red line is the exact solution.

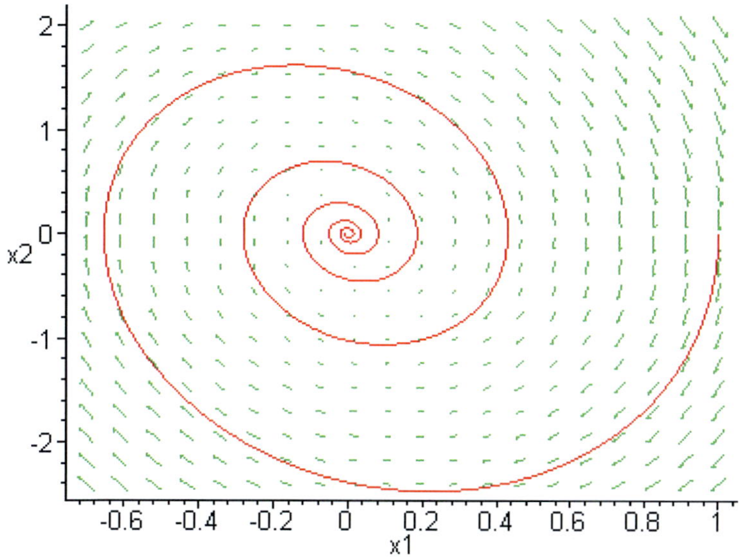

Figure 1.18. Phase diagram for simple damped oscillator with $\varepsilon = 0.5$

If we integrate and apply the initial conditions; then, the solution becomes:

$$x = x(0)\cos\left(t + \frac{3}{8}(x(0))^2 \varepsilon t\right) = \cos(t + \frac{3}{8}\varepsilon t)$$

Figure 1.19 shows a comparison of previous solution with a numerical solution.

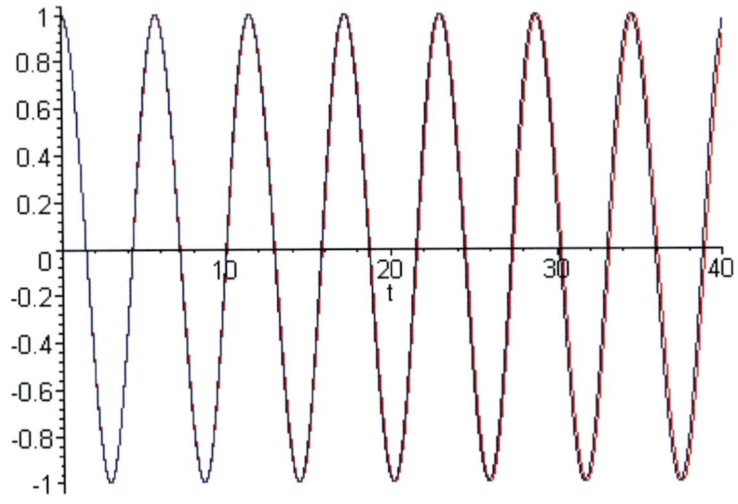

Figure 1.19. Time solution for the cubic non linear oscillator with $\varepsilon = 0.5$. Numerical result is in red and the approximate solution is in blue.

AVERAGE VALUE METHOD

The Two Scale Method cannot be applied to autonomous systems. Therefore, the Average Value Method is an alternative for the analysis of such systems. In 1934, Krylov and Bogoliubuv applied a form of the average value method to weak nonlinear system of the form:

$$\ddot{x} + \omega_0^2\, x + \varepsilon\, h(x, \dot{x}, t) = 0 \tag{1.47}$$

where h has a period T. In this way, forces with period T contrast with the "slow" contribution of the solution average. Therefore, in the same way as in the Two-Scale Method, the solution has the form:

$$x = r(t)\cos(\omega t + \phi(t)) \tag{1.48}$$

with the condition that

$$\dot{x} = -\omega\, r(t)\sin(\omega t + \phi(t)) \tag{1.49}$$

Thus, if we derive equation (1.48) with respect to time, we find that:

$$\begin{aligned} \dot{x} &= \dot{r}\cos(\omega t + \phi) - r\sin(\omega t + \phi)(\omega + \dot{\phi}) \\ &= \dot{r}\cos(\omega t + \phi) - \omega\, r\sin(\omega t + \phi) - r\,\dot{\phi}\sin(\omega t + \phi) \end{aligned} \tag{1.50}$$

and applying the statement of equation 1.49, then:

$$\dot{r}\cos(\omega t+\phi)-r\dot{\phi}\sin(\omega t+\phi)=0 \tag{1.51}$$

If we derive the conditioning equation (1.49) with respect to time, we find that:

$$\begin{aligned}\ddot{x} &= -\omega\dot{r}\sin(\omega t+\phi)-\omega r\cos(\omega t+\phi)(\omega+\dot{\phi})\\ &= -\omega\dot{r}\sin(\omega t+\phi)-\omega^2 r\cos(\omega t+\phi)-\dot{\phi}\,\omega r\cos(\omega t+\phi)\end{aligned} \tag{1.52}$$

Substituting these equations into the original differential equation, the system can be represented as:

$$\begin{aligned}&-\omega\dot{r}\sin(\omega t+\phi)+(\omega_0^2-\omega^2)r\cos(\omega t+\phi)\\ &\quad-\omega r\cos(\omega t+\phi)+\varepsilon h(r,\dot{r},t)=0\end{aligned} \tag{1.53}$$

If we simultaneously solve equations 1.51 and 1.53, we find that:

$$\dot{r}=\left(\frac{\varepsilon h}{\omega}+\frac{r}{\omega}(\omega_0^2-\omega^2)\cos(\omega t+\phi)\right)\sin(\omega t+\phi) \tag{1.54}$$

and

$$r\dot{\phi}=\left(\frac{\varepsilon h}{\omega}+\frac{r}{\omega}(\omega_0^2-\omega^2)\cos(\omega t+\phi)\right)\cos(\omega t+\phi) \tag{1.55}$$

if we define a new variable such that:

$$\Omega=\frac{\omega_0^2-\omega^2}{\varepsilon} \tag{1.56}$$

we can get the same results as Guckenheimer and Holmes, and Nayfeh and Mook. Guckenheimer and Holmes applied van der Pol's transform, while Nayfeh and Mook utilized the perturbation method.

$$\dot{r}=\frac{\varepsilon}{\omega}(h+r\Omega\cos(\omega t+\phi))\sin(\omega t+\phi) \tag{1.57}$$

$$r\dot{\phi}=\frac{\varepsilon}{\omega}(h+r\Omega\cos(\omega t+\phi))\cos(\omega t+\phi) \tag{1.58}$$

We find the final solution by the approximation of the *h* function with Fourier series. This part of the method is similar to the Two-Scale Method. Thus, equations 1.57 and 1.58 become:

$$\dot{r} = \frac{\varepsilon}{\omega}\langle h\sin(\omega t + \phi)\rangle \tag{1.59}$$

$$r\dot{\phi} = \frac{\varepsilon}{\omega}\left(\langle h\cos(\omega t + \phi)\rangle + \frac{1}{2}r\Omega\right) \tag{1.60}$$

To demonstrate the use of this method, we first solve the damped simple oscillator, as we did in previous sections. With this example, we can see that both methods provide the same results. The damped simple oscillator is represented by the following differential equation with initial conditions:

$$\begin{aligned} &\ddot{x} + 2\varepsilon\dot{x} + \omega_0^2 x = 0 \\ &x(0) = 1, \dot{x}(0) = 0 \end{aligned} \tag{1.61}$$

Using equation 1.48 and according to condition 1.49

$$\begin{aligned} x &= r(t)\cos(\omega t + \phi(t)) \\ \dot{x} &= -\omega r(t)\sin(\omega t + \phi(t)) \end{aligned} \tag{1.62}$$

The *h* function is defined as $h = 2\dot{x}$, thus:

$$h = -2\omega r\sin(\omega t + \phi)$$

Since $\Omega = 0$, and applying equation 1.59 and 1.60, the set of equations become:

$$\dot{r} = \frac{\varepsilon}{\omega}\langle h\sin(\omega t + \phi)\rangle = -\varepsilon r$$

$$r\dot{\phi} = \frac{\varepsilon}{\omega}\langle h\cos(\omega t + \phi)\rangle = 0$$

Solving for *r* and ϕ:

$$r = c_1\mathbf{e}^{-\varepsilon t}$$
$$\phi = c_2$$

Moreover, applying the initial conditions:

$$r(0) \approx \sqrt{x(0)^2 + \dot{x}(0)^2} = 1 \Rightarrow c_1 = 1$$

$$\phi(0) \approx \tan^{-1}\left(\frac{\dot{x}(0)}{x(0)}\right) = 0 \Rightarrow c_2 = 0$$

In this way, we obtain the same expression as with the previous method.

$$x = \mathbf{e}^{-\varepsilon t}\cos(\omega t)$$

Another advantage of the Average Value Method is its ability to determine the amplitude response as a function of the excitation frequency ω. With this feature, we can analyze the behavior of systems such as those modeled with the Duffing's equation:

$$\ddot{x} + \omega_0^2 x + \varepsilon\left(-\gamma\cos\omega t + \delta\dot{x} + \alpha x^3\right) = 0$$

Following the previous description, we assume that the solution will behave as:

$$\begin{aligned} x &= r(t)\cos(\omega t + \phi(t)) \\ \dot{x} &= -\omega r(t)\sin(\omega t + \phi(t)) \end{aligned}$$

Thus, function h is expressed as:

$$\begin{aligned} h &= -\gamma\cos\omega t + \delta\dot{x} + \alpha x^3 \\ &= -\gamma\cos\omega t - \omega\delta r\sin(\omega t + \phi) + \alpha r^3\cos^3(\omega t + \phi) \end{aligned}$$

Substituting h into equation 1.49, the set of differential equations is defined as:

$$\begin{aligned} \dot{r} &= \left\langle\left(\frac{\varepsilon h}{\omega} + \frac{r}{\omega}\left(\omega_0^2 - \omega^2\right)\cos(\omega t + \phi)\right)\sin(\omega t + \phi)\right\rangle \\ &= \frac{\varepsilon}{\omega}\left\langle\left(-\gamma\cos\omega t - \omega\delta r\sin(\omega t + \phi) + \alpha r^3\cos^3(\omega t + \phi)\right)\sin(\omega t + \phi)\right\rangle \\ &= \frac{\varepsilon}{\omega}\left(\left\langle -\gamma\cos\omega t\left(\sin\omega t\cos\phi + \cos\omega t\sin\phi\right)\right\rangle - \frac{1}{2}\omega\delta r + 0\right) \\ &= \frac{\varepsilon}{2\omega}\left(-\gamma\sin\phi - \omega\delta r\right) \end{aligned}$$

Similarly, substituting h into equation 1.60, we find that:

$$r\dot{\phi} = \left\langle \left(\frac{\varepsilon h}{\omega} + \frac{r}{\omega}\left(\omega_0^2 - \omega^2\right)\cos(\omega t + \phi)\right)\cos(\omega t + \phi)\right\rangle$$

$$= \frac{\varepsilon}{\omega}\left(-\gamma\cos\omega t - \omega\delta r\sin(\omega t + \phi) + \alpha r^3\cos^3(\omega t + \phi)\right)\cos(\omega t + \phi) + \frac{1}{2\omega}r\left(\omega_0^2 - \omega^2\right)$$

$$= \frac{\varepsilon}{\omega}\left[-\frac{1}{2}\gamma\cos\phi + \frac{3}{8}\alpha r^3\right] + \frac{1}{2\omega}r\left(\omega_0^2 - \omega^2\right)$$

$$= \frac{\varepsilon}{2\omega}\left(-\gamma\cos\phi + \frac{3}{4}\alpha r^3 + r\Omega\right)$$

When the system behaves in a steady state, there is a singular solution. This solution is also known as "fixed points". In this case, the fixed points are located at:

$$\dot{r} = 0$$

$$\dot{\phi} = 0$$

Then, the system can be represented as:

$$-\omega\delta r = \gamma\sin\phi$$

$$\frac{3}{4}\alpha r^3 + r\Omega = \gamma\cos\phi$$

Combining these two equations:

$$(-\omega\delta r)^2 + \left(\frac{3}{4}\alpha r^3 + r\Omega\right)^2 = \gamma^2$$

Solving for Ω

$$\Omega = \frac{-3\alpha r^3 \pm 4\sqrt{\gamma^2 - (\omega\delta r)^2}}{4r}$$

$$= \frac{\omega_0}{\varepsilon}\left(1 - \frac{\omega}{\omega_0}\right)$$

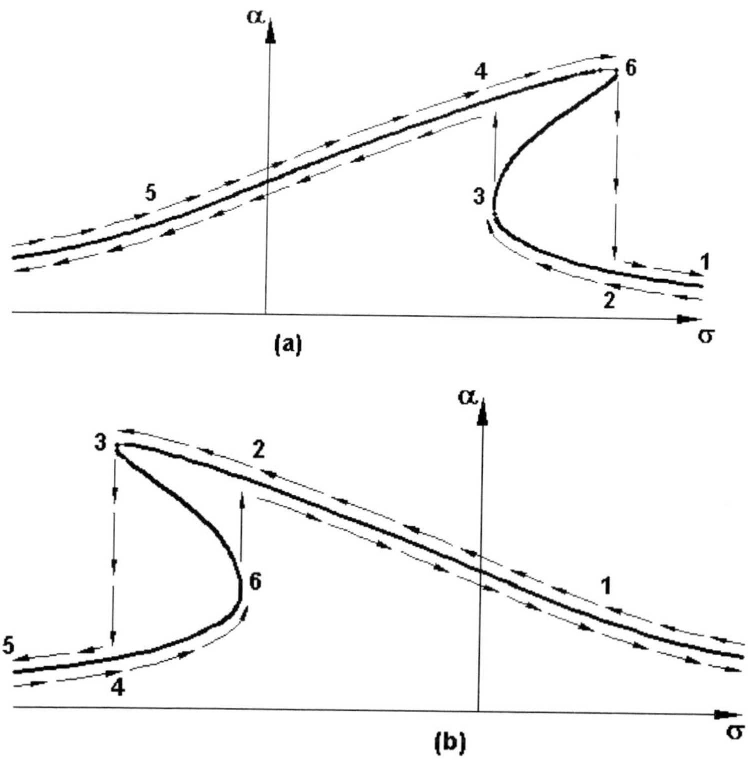

Figure 1.20. Jump phenomena for primary resonance of the Duffing equation: (a) $\alpha > 0$; (b) $\alpha < 0$.

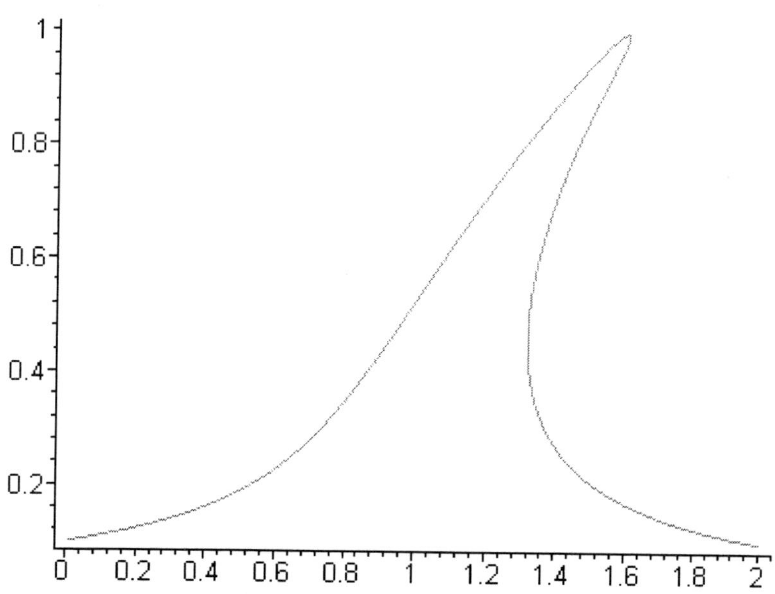

Figure 1.22. Solution for the Duffing oscillator with $\varepsilon = 0.5$ y . $\alpha > 0$.

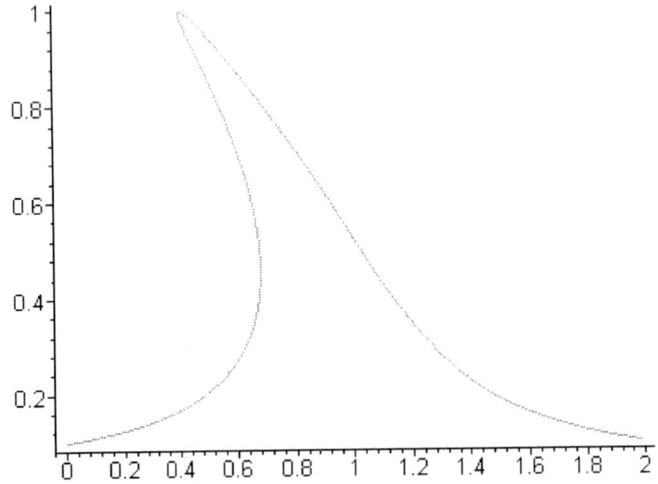

Figure 1.22. Solution for the Duffing oscillator with $\varepsilon = 0.5$ y . $\alpha < 0$.

In general, γ, δ and ω_o, are constant; thus, if we solve for different values of ω, we can see that there is a range where there are two solutions, these solution correspond to an instable state that is identified as "jump". Jump phenomenon occurs in many mechanical systems, in the following chapters, we will present different systems that show such behavior. In those systems, if we increase the speed, the excitation forces will increase their frequency and vibrating amplitude will also increase. In a linear system, the amplitude will depend upon the damping factor δ and it will have its maximum amplitud when $\omega = \omega_o$; but in this case, the amplitude continues increasing until $\omega > 1.6\omega_o$, at this point the amplitude drops drastically and remains low up to $\omega > 2\omega_o$. In order to identify the jump phenomenon, we have to reduce the speed. The amplitude response is totally different, it maintains a low value until $\omega = 1.3\omega_o$, at this point the amplitude increases drastically, and it continues reducing its value monotonically.

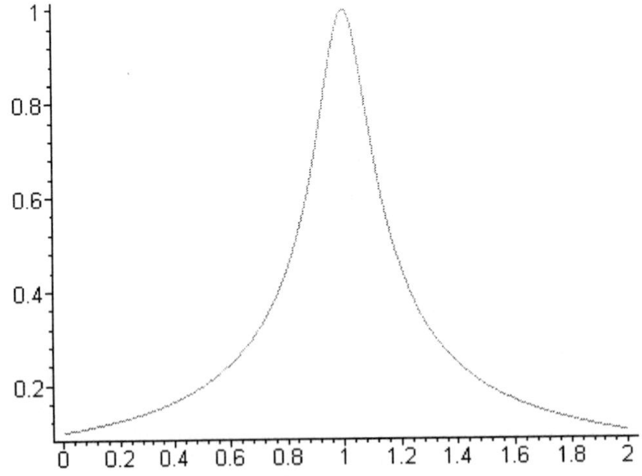

Figure 1.23. Solution for the Duffing oscillator with $\varepsilon = 0.5$ y . $\alpha = 0$.

α plays an important rule in the system response. It determines the amount of "curvature" of the graph. As α increases, the curve will have a larger bent. Even more, the sign of α determines the orientation of the bent, if $\alpha > 0$ the bent is to the left, and if $\alpha < 0$, the bent is to the right. The physical interpretation of the shape of the curve is: To the left is known as softening stiffness and to the right hardening stiffness.

POINCARE MAPS

We include this concept as a reference, it is an important technique for the analysis of nonlinear systems, but in the case of mechanical systems, it has limited contributions. They can illustrate the trajectory flow of a periodic orbit. The basis for its construction is a three dimensional plot of the state response. The plane of this graph is formed with the two state variables (usually the normalized displacement and the normalized velocity) and the third axis is the time variable. In this way, elliptic orbits growth and represent the solution as surfaces in the state solution domain.

The Pincaré maps are constructed by cutting the surface with intersecting planes. Thus, we get a set of points at the intersections of the orbits and the plane. The nature of the system determines the distribution of the set of points along the plan, and the analysis of them shows the existence of instabilities, transient responses or nonlinear behaviors. Therefore, the Poincaré map simplifies a closed orbit problem into a fixed point's map, which presumes to be a simplification, but it creates a new difficulty that is the definition of a formula to define any point in the map. The easiest way of applying Poincaré maps is graphically.

For example, if we have a system defined by:

$$\begin{aligned}\dot{x}_1 &= x_1 - x_2 - x_1\left(x_1^2 + x_2^2\right)\\ \dot{x}_2 &= x_1 + x_2 - x_1\left(x_1^2 + x_2^2\right)\\ \dot{x}_3 &= \mu x_3^2\end{aligned} \tag{1.63}$$

with fixed points:

$$X = \begin{cases}(0,0,0)\\(1,0,0)\\(-1,0,0)\end{cases}$$

Figure 1.24 shows the space surface of the state solution. If we intersect a *xz* plane at $y = 0$, we reduces the problem from N to a N *-1* dimension.

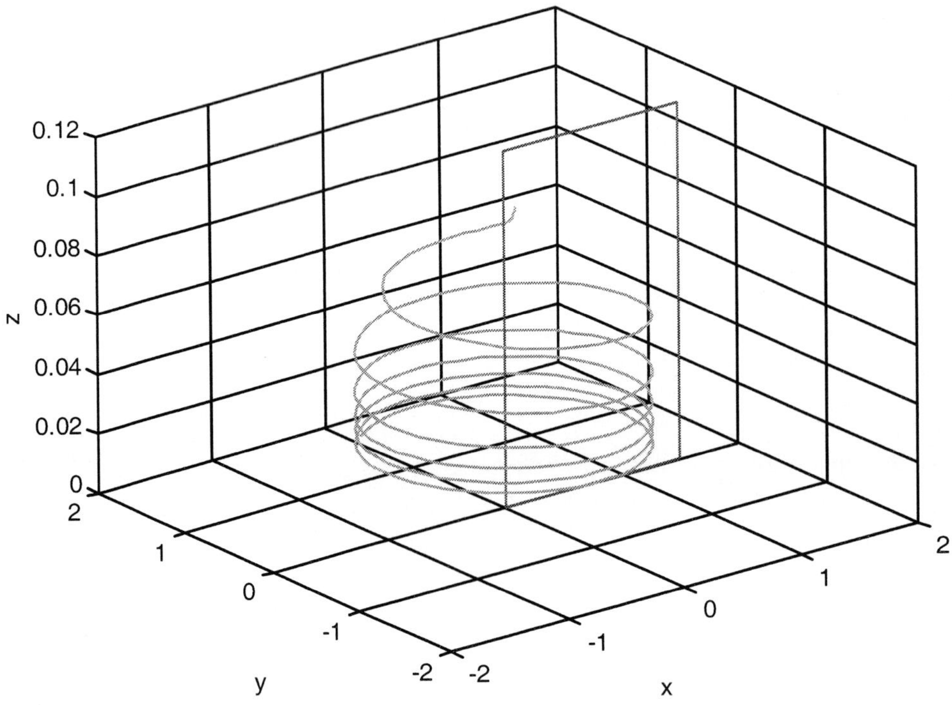

Figura 1.24. Representation of a Poincaré Map for the system (1.63).

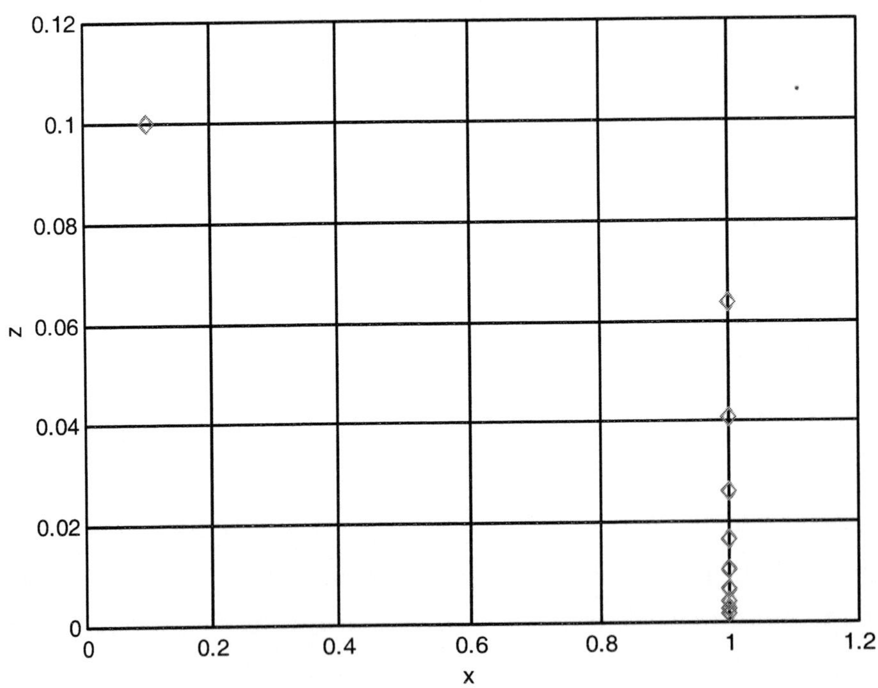

Figure 1.25. Poincaré map for the system (1.63).

Figure 1.25 shows on of the Poincaré maps. In this map, we can see the points distribution and how the solution varies with time. At point *(1,0,0)* we identify a limit cycle followed by a stable spiral.

Although Poincaré maps seem to be an attractive solution, in many cases it is almost imposible to obtain an analytical expression for the reduced solution. Even more, this analytical solution could be more complicated than the original system solution. Another possibility is through a numerical solution, in which case, the stability of the solution should not interfere with the stability analysis.

APPENDIX

$h(\theta)$ can be obtained from the Fourier coefficients with the following identity:

$$\frac{1}{2\pi}\int_0^{2\pi} h(\theta)f(\theta)d\theta \equiv \langle h\, f(\theta)\rangle$$

The most common averages that appear in solving a system include the following:

$$\langle \cos \rangle = 0$$

$$\langle \sin \rangle = 0$$

$$\langle \sin\cos \rangle = 0$$

$$\langle \cos^2 \rangle = \frac{1}{2}$$

$$\langle \sin^2 \rangle = \frac{1}{2}$$

$$\langle \sin^2\cos^2 \rangle = \frac{1}{8}$$

$$\langle \cos^3 \rangle = 0$$

$$\langle \sin^3 \rangle = 0$$

$$\langle \sin^3\cos^3 \rangle = 0$$

$$\langle \cos^4 \rangle = \frac{3}{8}$$

$$\langle \sin^4 \rangle = \frac{3}{8}$$

$$\langle \sin^4\cos^4 \rangle = \frac{3}{128}$$

Chapter 2

MECHANICAL SYSTEMS WITH STIFFNESS DISCONTINUITIES

INTRODUCTION

In the first chapter we presented basic ideas of the analysis of nonlinear systems and their vibrating response. These nonlinear vibration concepts were shown in a very general perspective, without describing the relationship between the mathematical terms and the physical characteristics of mechanic systems. It is well known that most mechanical systems present linear dynamic responses. And even more, in most cases their transient response has no practical implications. From a linear analysis point of view, the best way of analyzing those systems is through the application of Fourier analysis. With the aid of the Fast Fourier Transform (FFT) and modern accelerometers, vibration analysis has become a powerful tool for industrial applications. Nowadays, predictive maintenance systems relay basically on vibration analysis, they capture the vibration signal of a machine with an accelerometer and they produce a frequency spectrum using the FFT; from the spectrum, experts can identified the vibration source and they can diagnose the machine. But this procedure is unable to identify vibrations which come out from a nonlinear source. What has been reported as the response of such sources is the result of the numerical process of the nonlinear signal. Therefore, a better understanding of the relation between the mechanical design and the nonlinear effects is needed. Thus, a suitable model for each system can be written and the solution of this model can give better reference for the analysis of field problems.

In this chapter we describe systems whose stiffness present discontinuities. This is one most recurrent parameter that characterized a nonlinear dynamic system. But, it is no a simple task to link the mechanical design to the stiffness parameter. Even more, we have identified a number of mechanical systems that can be model with a simplified expression. In many of them, their equation of motion resembles a Duffing's equation; therefore, the nonlinear vibration theories can be applied directly. In this way, we have found that a stiffness discontinuity can be approximated as a cubic polynomial. Common cases are gear boxes, sliding bearings, rotor rubbing and nonlinear bearings, mechanical backlash and many others.

Another particular difference of the stiffness discontinuity is that it can depend on position, frequency or time. Thus, the analysis produces a totally different output. For that reason, other analysis tools are required, such as phase diagrams or time-frequency maps.

Some of these analyses are included in this chapter and they are also described in Chapter 5. Future generations of monitoring systems will include time-frequency maps and phase diagrams as analysis tools.

Mechanical systems present three types of stiffness discontinuities, they depend on the stiffness function; thus, gearboxes and rotor rubbing varies their stiffness as a function of their position, In the case of bearings, their stiffness function depends on frequency, and backlash (except on the case of gears) depends on time.

GEARBOXES

They represent one of the most studied mechanical systems. Its design has evolved from the ancient machines, and they have reached one of the most specialized design. Although their purpose is to transmit power between a power source and a specific device, by themselves they compound different fields of knowledge. As a complete system, they content gears or teeth wheels, shafts, bearings, rolling bearings, lubrication pumps, tubes, valves and other devices such as heat exchangers and others. Therefore, all these individual elements have gone through a development process by themselves, but as an integrated system they have challenge engineers with high interesting problems. The one that brings our attention in this book are gear vibrations, that are always undesirable, and they also generate noise. The dominant cause of gear noise is the Transmission Error, it is referred as the deviation from a perfect motion between the driver and the driven gears. And it is the combination of different gear variations, such as non perfect tooth profile, pitch errors, elastic deformations, backlash, etc. The simplest type of noise is a steady note which may have a harmonic content at gear mesh frequency. This frequency is normally modulated by the rotating frequency. Modulated noise is often described as buzzing sound. In general, gears show a frequency modulated spectrum with a distinguished mesh frequency and side bands space at the shaft rotating frequency. Other noises are associated with pitch errors. They are described as scrunching, grating, grouching, etc. They contain a wide range of frequencies that are a lot higher than the rotating frequency. White noise can also be present and it may be associated with loss of contact between the teeth.

Gear box vibration is a typical nonlinear vibration phenomenon. Its nonlinear behavior comes from the discontinuities in the stiffness of the system, which comes from the combination of two teeth acting in conjunction. Thus, the stiffness of a gear pair varies with the angular position, except in very specific gear designs. One of the main features of gear pair stiffness is that it changes drastically as a function of the number of teeth in simultaneous contact. Ideally, a pair of gears transmits motion at a constant speed (Figure 2.1). This feature comes from the involute shape of their teeth, but in reality there are other effects that affect this property, mainly their elastic deformation deviates the actual trajectory and the stiffness of the gear action becomes a combination of the relative position and the number of teeth in contact. It has a periodic function but it can be approximated as a rectangular periodic function. Figure 2.4 shows a graphical approximation of the stiffness of the system as a function of the angular position. It is clear that it is no constant, and it can be approximated as a rectangular periodic function.

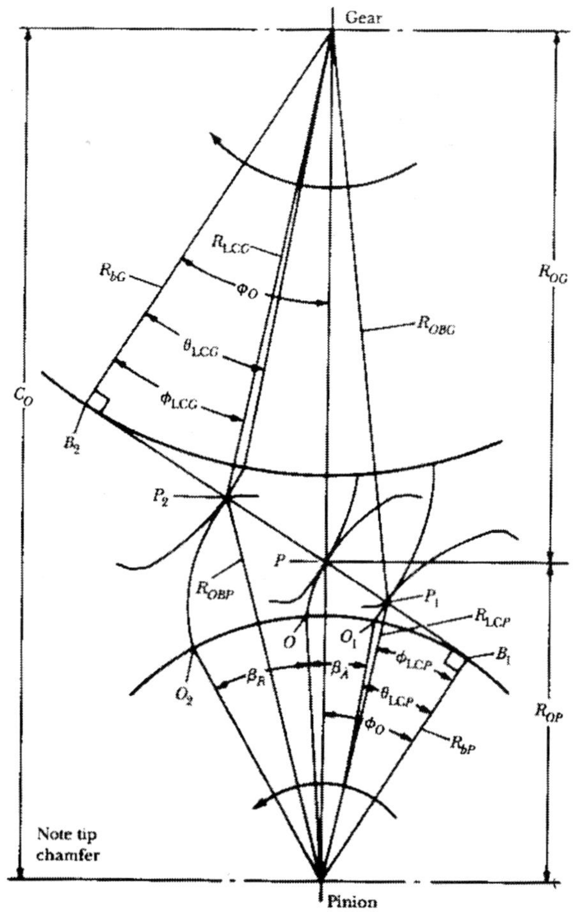

Figure 2.1. Gear conjugated motion.

Contact between teeth takes place along the line of action; that is, the distance where matching gears are in contact, limited by the outside diameters of both gears. Gear action is divided in two: from the beginning of contact until they reach the pitch diameter, and from the pitch diameter until the end of contact. These two sections have significant importance because the sliding velocity is different, in the first case it is negative, thus the oil film is reduced, whereas in the second case, the sliding velocity is positive and it brings more oil to the contact area. It is remarkable that the variation in oil film causes variations in the damping coefficient as the teeth transmit motion.

One of the important parameters that affect gear dynamics is its contact ration. Contact ratio can be derived from a spur gear action, and then extrapolated to other geometries, its demonstration is out of the scope of this book, but the general formula is:

$$m_p = \frac{\sqrt{\left(R_{OBP}^2 - R_{bP}^2\right)} + \sqrt{\left(R_{OBG}^2 - R_{bG}^2\right)} - C_O \sin(\theta_O)}{P_b} \tag{2.1}$$

During gear action, more than one pair of teeth is in contact. The contact ration defines the percentage of time that more than one pair of teeth are in contact. For example, if the gear ration is 1.43, it means that 43% of the time there are two pair of gears in contact and 57% of the time only one pair of teeth transmits the torque. This is the dominant effect on the gear mesh stiffness.

Gear dynamics can be predicted with a single degree of freedom model or a multiple degrees of freedom model. Al depends upon the problem under study. To increase the accuracy there are several alternatives: either increase the number of degrees of freedom or add complex parameter that consider nonlinear terms, defects, time variation stiffness of the gear mesh, bearings, housing stiffness, and others less common. Gear dynamic models may also include lateral and torsional vibrations, both cases are always present in a gear box, but torsional vibrations are more complicated to measure in operating equipment. In this book we developed a general model for a pair of gears, we define the excitation forces in a way that it includes geometry and fabrication errors; the gear mesh stiffness is defined as a discontinuous function and its variations are included into the system stiffness. Although impulse impacts due to oil film fluctuations are present, we ignore them because we want to illustrate the nonlinear discontinuities in the gear mesh stiffness.

In most gear pair systems, torsional motion is couple by the gear pair stiffness; therefore a two degree of freedom model will reflect accurately most practical applications. If it is necessary to include other effects, increasing the degrees of freedom could accommodate other compliances that are present in the system.

Many researches and engineers have developed a significant number of gear dynamic models. Most of them are been developed for the prediction of noise and vibrations, and they have demonstrated that gear vibrations are highly nonlinear. In this chapter we presented one of the most commonly use model that is widely accepted. It has been demonstrated (Velex, 1996, Smith 1983) that a simplified lumped-mass model is adequate for small transmissions. Figure 2.2 shows a simple dynamic model of a gear pair. There are many representations of this system, but this can identify most of the common vibration and noise problems. The only ones that it can not reproduce are those associated with backlash.

In the case of gear boxes with several speed reductions and light casings more sophisticated models are needed. On the other hand, this simplified model illustrates how a gear reduction can be analyzed as a system with stiffness discontinuities, and it allows us the identification of this characteristic in the vibration response.

The model is compound by two masses M_1 and M_2,(pinion, gear and their shafts) with their corresponding rotational inertias J_1 and J_2, pitch diameters r_1 and r_2, torsional stiffness K_1 and K_2, lateral stiffness S_1 and S_2, teeth stiffness S_t, and gear force P.

Applying Lagrange's equations, where the generalized displacements are x_1, x_2, θ_1 and θ_2, a system of four differential equations is found.

$$
\begin{aligned}
m_1\ddot{x}_1 &= (S_1 + S_t)x_1 - S_t x_2 + S_t r_1\theta_1 - S_t r_2\theta_2 = P(t) \\
m_2\ddot{x}_2 &= (S_t + S_2)x_2 - S_t x_1 - S_t r_1\theta_1 + S_t r_2\theta_2 = 0 \\
J_1\ddot{\theta}_1 &= (S_1 r_1 + 2S_t r_1)x_1 - 2S_t r_1 x_2 + 2S_t r_1^2\theta_2 - 2S_t r_1 r_2\theta_2 = 0 \\
J_2\ddot{\theta}_2 &= (S_2 r_2 - 2S_t r_2)x_2 + 2S_t r_2 x_1 + 2S_t r_1 r_2\theta_1 - 2S_t r_2^2\theta_2 = 0
\end{aligned}
\tag{2.2}
$$

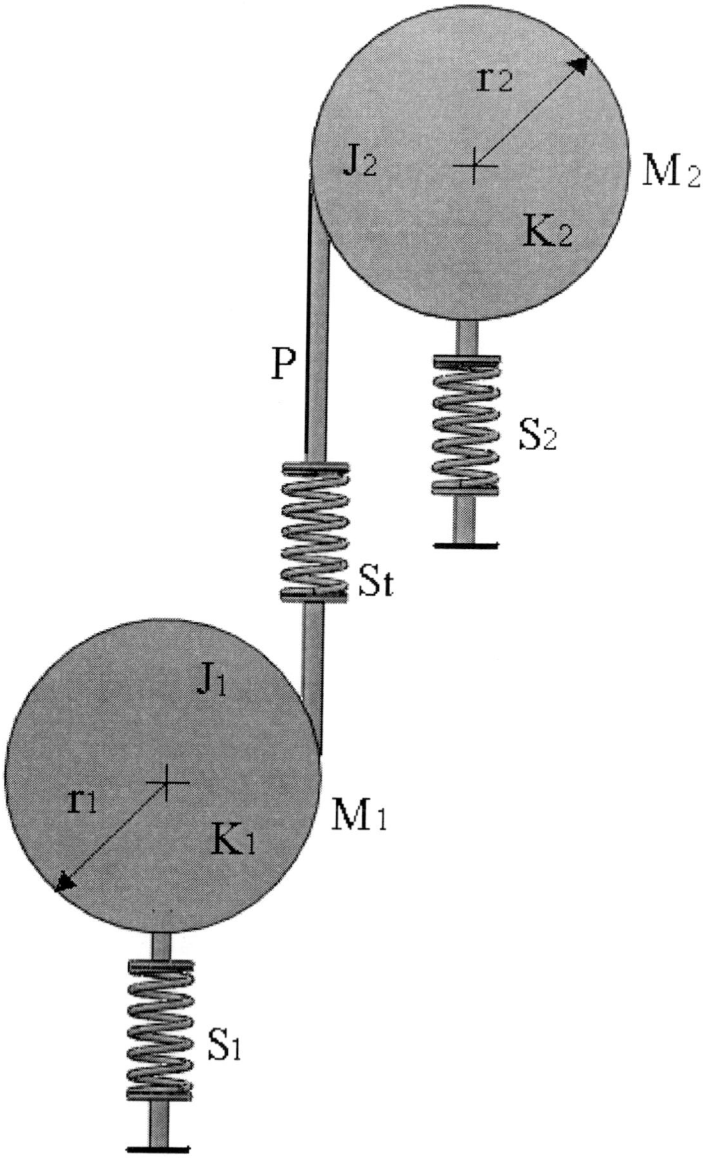

Figure 2.2. Simplified gear model.

The external force *P(t)* accounts for gear profile defects such as manufacturing errors, lead angle error, profile, error and so on. It also includes backlash effects, but since this is a specific topic it is treated later. The main element that causes the dynamic discontinuity is teeth stiffness; it is estimated from the gear action function and the contact ratio. It is compound from the stiffness of each tooth that is in contact, and the individual stiffness is determined considering that a tooth behaves as cantilever beam (Figure 2.3)

Then, from the beam theory, the bending stress is:

$$\sigma = \frac{Mc}{I} = \frac{12W_t\,L\,h}{2\,b\,h^3} = 6\frac{W_t\,L}{b\,h^3} \tag{2.3}$$

where b is the face with, W_t is the transmitted force, L is the height of Lewis' parabola and h its width. Considering the beam's maximum deflection δ:

$$\delta = \frac{W_t L^3}{3\,E\,I} \tag{2.4}$$

Thus, tooth stiffness can be approximated as:

$$k = \frac{3EI}{L^3} = \frac{1}{4}E\,b\left(\frac{h}{L}\right)^3 \tag{2.5}$$

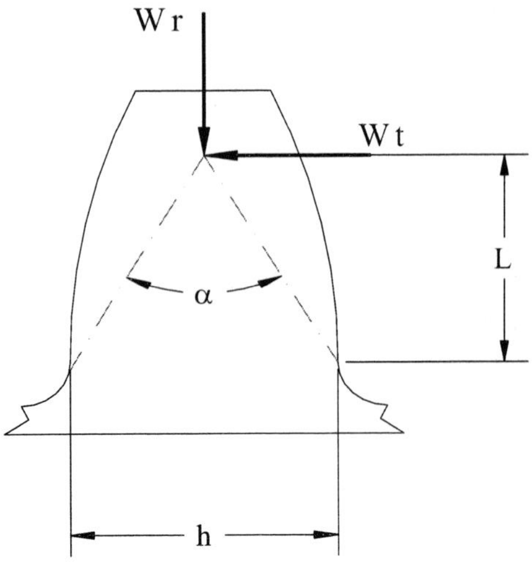

Figure 2.3. Forces acting on a tooth.

From Figure 2.3, it is clear that $tan(\alpha/2) = h/(2L)$, and in most case α=30°, then:

$$k = 2E\,b\left(\tan\left(\frac{\alpha}{2}\right)\right)^3 \tag{2.6}$$

As stated before, S_t combines the effect of teeth action, and it depends directly to the contact gear ratio m_p. Figure 2.4 shows the gear mesh stiffness function, it can be expressed as:

$$S_t = \begin{cases} 2Eb\tan\left(\dfrac{\alpha}{2}\right)^3 (n+1), \text{ para } 0 < \theta < \dfrac{2\pi}{N}(m_p - n) \\ 2Eb\tan\left(\dfrac{\alpha}{2}\right)^3 (n+1)?, \text{ para } \dfrac{2\pi}{N}(m_p - n) < \theta < \dfrac{2\pi}{N} \end{cases} \tag{2.7}$$

where N is the gear teeth number and n is the minimum teeth in contact.

It is clear that the gear mesh stiffness function is a periodic function, thus it can be expanded as a Fourier's series, thus

$$S_t = A_0 - A_1\theta + A_2\theta^2 - A_3\theta^3 \tag{2.8}$$

Coefficients A_i are only function of m_p, and θ represents gear rotation angle.

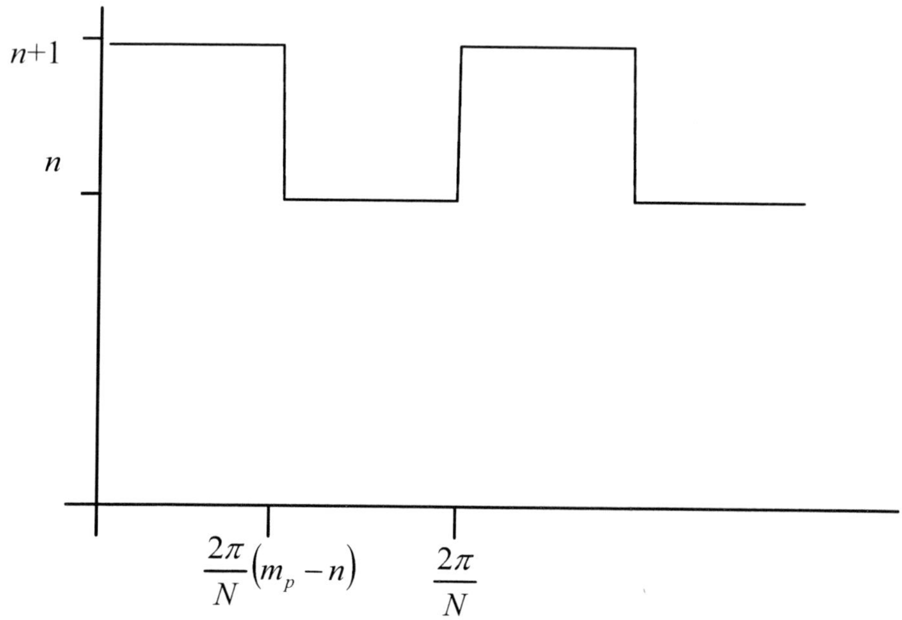

Figure 2.4. Gear mesh stiffness as a function of gear position.

The solution of the simple gear model requires the definition of the transmitted force *P(t)*. Ideally, it should be constant, and equal to the transmitted torque, divided by the gear radius. But, in real life this is not the case; a general approximation of the gear force includes, run out effects, gear profile errors, pitch errors, and manufacturing errors. Thus, the external gear force has at least these terms:

Run out effect (F_r) is a sinusoidal function with a $2\pi/\omega$ period and amplitude of the gear eccentricity.

Pitch error (f_p) is a short wave length function with a period of $2\pi/\omega N$, and the accumulated pitch effort (F_p) is a long wave function with a period of $2\pi/\omega$.

The profile angle error is another source of vibration and it can be represented as long period function: $F_{\varphi} = \varepsilon \tan(\omega t)$

Other effects can also be included but in general they are present only in special cases that are out of the scope of this chapter. Finally, the external forces can be combined as:

$$P(t) = (F_p + F_r)\sin(\omega t) + (f_p + F_{\varphi})\sin(N\omega t) \tag{2.9}$$

where ω is the gear angular velocity.

Gears also have some damping, it is cause by friction forces during teeth contact, and they are determined using modal damping. On the other hand, friction depends on the relative velocity during gear action and on the oil film thickness. The sliding velocity is calculated as:

$$v_{SP} = (\rho_P \omega_P - \rho_G \omega_g) \tag{2.10}$$

and the rolling radius is calculated as

$$\rho_P = R_{bP} \tan(\alpha) \tag{2.11}$$

where α is the pressure angle at the radius of contact. It is important to remark that the pressure angle varies along the line of action, and its reference value is defined at pitch diameter. Thus, the sliding velocity depends upon the position of the teeth along the line of action.

Damping coefficients should account for less than 3% of the damping ratios of the vibration modes.

$$C_{gm} = \alpha M_{gm} + \beta K_{gm} \tag{2.12}$$

It is clear that, even for the simplest gear model, a gear box will develop chaotic and periodic motions. The quadratic and cubic terms of the gear mesh stiffness are relatively high compared to the linear terms; then, the system can be analyzed as a Duffing's system.

Next is an example of a gear box vibration. It is a transmission with two spur gear and the initial conditions were set as $x_1=0$ and $X_1=1$ (see solution methods Chapter 5). Figure 2.5 shows the time response of the pinion. It is noticeable that the response is periodic but not sinusoidal; this is a typical characteristic of a harmonic distortion that will be discussed latter.

As discuss in previous chapters, the phase diagram is a clear indication of the nonlinear behavior of a gear pair, it creates to attracting poles, as shown in Figure 2.6. Since the gear system can be modeled as Duffing's system, another characteristic of its response is that it frequency diagram shows the jump phenomenon.

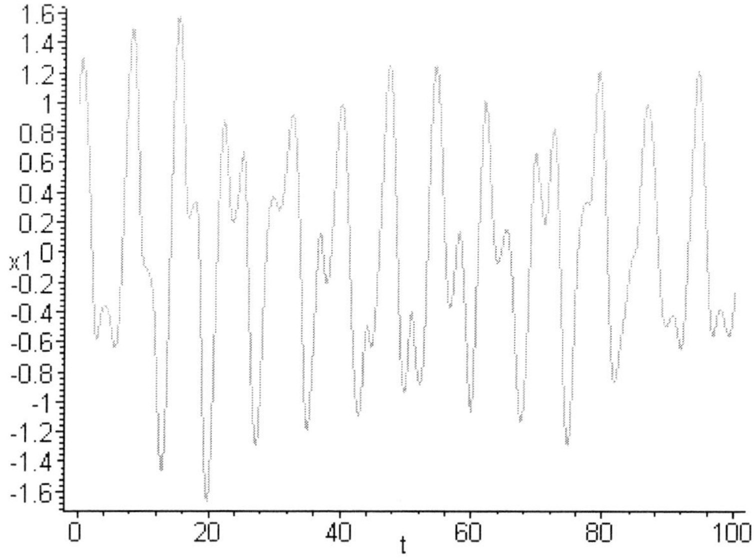

Figure 2.5. Time response, with initial conditions $x_1 = 0$, $X_1 = 1$.

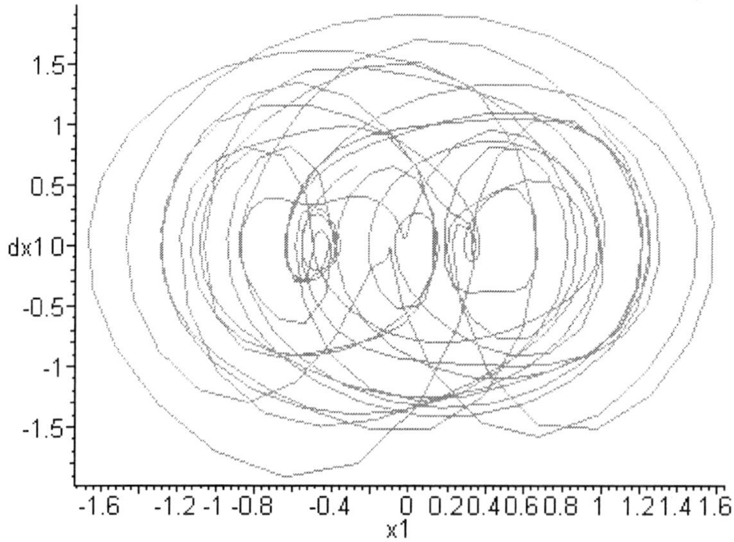

Figure 2.6. Phase diagram for the same initial conditions $x_1 = 0$, $X_1 = 1$.

Experimental results have also show this type of behavior. Laboratory test have also found jump phenomenon. Figure 2.7 illustrates how at different frequencies a gear box developed jump.

It is important to know that there are two types of vibrations in a gear box. One comes from the external forces, and they have a well described form typical of a modulated signal: Short wave signal at gear speed frequency over a long wave signal at a frequency equal the speed frequency times the number of teeth (known as gear frequency). These two signals are easily identified with any vibration measuring system. Accuracy manufacturing is the only way of reducing the effects of this type of vibration. The other vibration comes from the nonlinear vibration. In this case, the source of it is the discontinuity on the gear mesh stiffness plus other transmission errors. Its main parameter is the contact ratio, and it is intrinsic to the gear pair design, therefore, the only way of minimizing this vibration is through the designing of the gear pair in a way that the total contact ratio is an entire number and it is at least two.

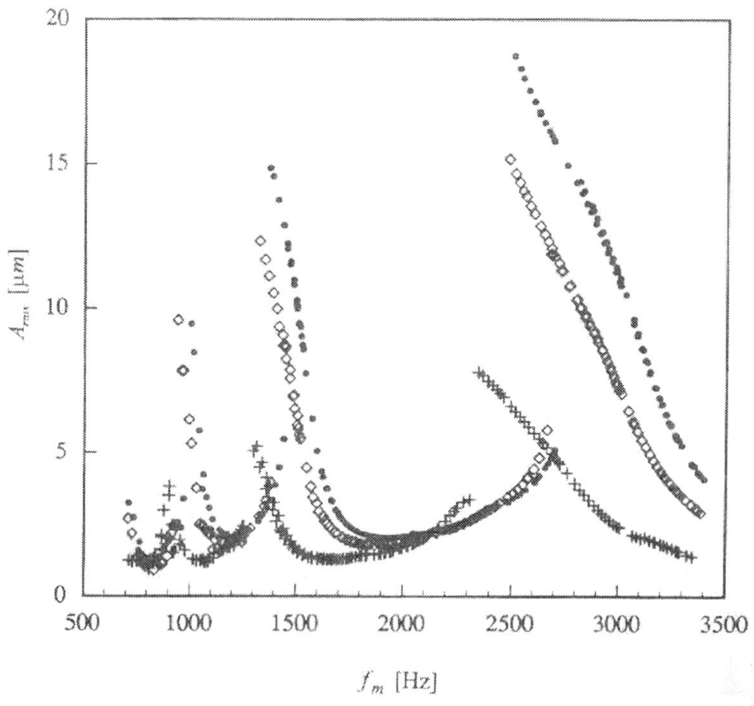

Figure 2.7. Experimental data.

BEARINGS

Bearings support rotors and provide load-carrying capacity. They are basically to types: Rolling bearings and sliding bearings. In a sliding bearing, there is a fluid between the rotor and the support surface. In rolling bearings, the surfaces are separated by balls or rollers. Another type is the magnetic bearing. Although most of the bearings show a nonlinear behavior, only some of them present stiffness discontinuities

New journal bearings, such as gas foil bearings show also nonlinear stiffness of the form

$$F = k_1 r + k_2 r^2 + k_3 r^3 \tag{2.13}$$

where r is the radial displacement of shaft. But this type of nonlinearity shows no discontinuity in the stiffness, and the dynamic response will show high vibrations at subsynchronous frequencies, rather than at higher frequencies. These types of stiffness nonlinearities are continuous in nature, thus they are out of the scope of the book. We only make a reference to clarify that not all nonlinear stiffness can be treated as a discontinuous systems, therefore their analysis and identification in field problems is also different. In contrast, ball bearings and sliding bearings show stiffness discontinuities, and their nonlinear vibrations have a different response compare to journal bearings. It is noticeable that equation 2.13 is a cubic equation on the force, but a quadratic equation on the stiffness.

A typical ball-bearing system consists of five contact parts: the shaft, the inner ring, the rolling elements, outer ring and the housing. The deformation of each part will influence the load distribution and, in turn, the service life of the bearing. It is well known that classical calculation methods cannot predict accurately load distributions inside the bearing. Ball bearings (Figure 2.8) are very stiff compared with sliding bearings, their stiffness can be approximating as a set of individual springs; where the number of springs supporting the shaft varies with the angular position of the shaft. This variation depends upon the kinematics of the ball o roller as it moves around the shaft. Thus, the ratio of rotation of the ball as a function of shaft's rotation is determined as

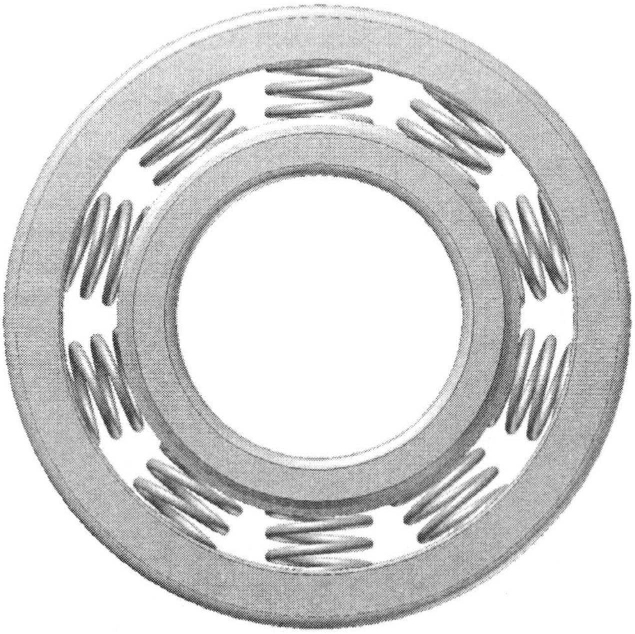

Figure 2.8. Schematic representation of a roller bearing.

The fundamental principle of a rolling bearing is that the ball or roller translates around the shaft, eliminating must of the friction; then the ball's angular translation is found as (*D* is the pitch diameter and *d* is the roller diameter)

$$\phi = \frac{d\cos(\omega_b t)}{D + \frac{d}{2}}\cos(\alpha) \tag{2.15}$$

The number of balls, or rolls in contact are determined from Figure (2.9). The nonlinear characteristic of the rolling bearing is the ball-track deformation. The ball-track stiffness is calculated with Hertz equation. Since the balls translate around the shaft, the number of balls supporting the load varies with the angular position of the shaft; this translation effect modifies the overall stiffness of the bearing. Although this variation may be small, it creates a nonlinear vibration, which turns out to be relatively difficult to identify in field problems.

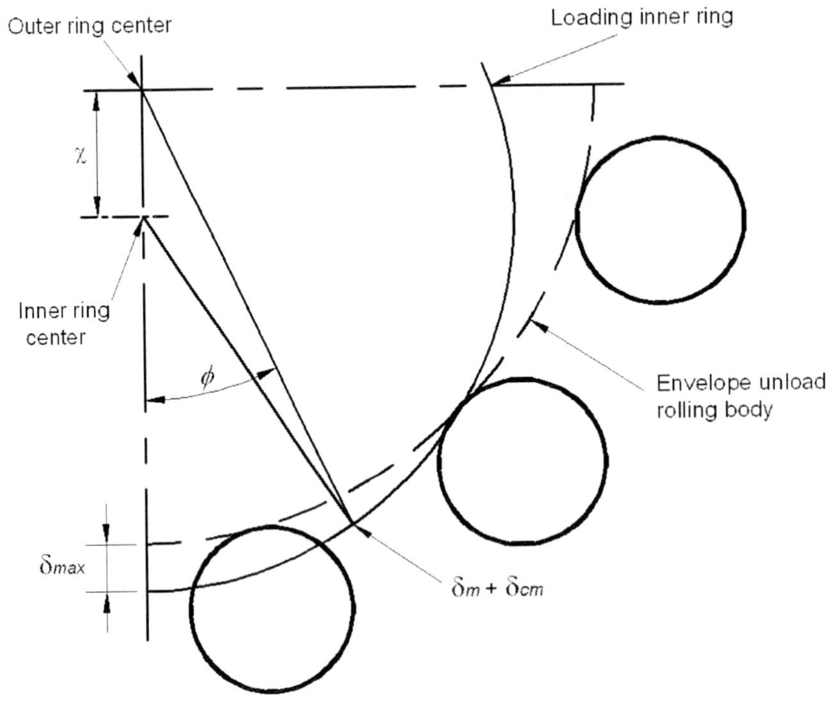

Figure 2.9. Radial displacement of a shaft mounted on ball bearings.

$$\omega_b = \frac{D\omega_s}{d} \tag{2.14}$$

Rolling bearings generate transient vibrations due to stiffness nonlinearities and structural defects. There are four external sources of vibration, two of them are associated with the angular velocity of the ball ω_b and their angular translation ω_ϕ. The other two frequencies are related to structural defects on the inner and outer tracks. These external frequencies excite the nonlinear terms and with the transient vibrations difficult its analysis with the Fourier's transform.

The stiffness of the ball as a function of the deformation is almost constant:

$$P_i = E\sqrt{\frac{Dd}{(D-d)}\left(\frac{\delta_i}{\alpha_H}\right)^3} \tag{2.16}$$

The non-linear effect comes from the combination relative deformation of the balls as they roll around the shaft. The rolling bearing can be modeled as a mass-spring system.

$$\delta_i = \delta_{\max} \cos\left[\frac{\pi}{2} - i\left(\phi + \frac{2(i-1)\pi}{N}\right)\right] \tag{2.17}$$

The spring stiffness is determined from Figure 2.10. Similarly as the gear mesh stiffness, rolling bearings exhibits a periodic function, thus it can be expanded as a Fourier's series:

$$k_x = a_0 + a_1\phi + a_2\phi^2 + a_3\phi^3 \tag{2.18}$$

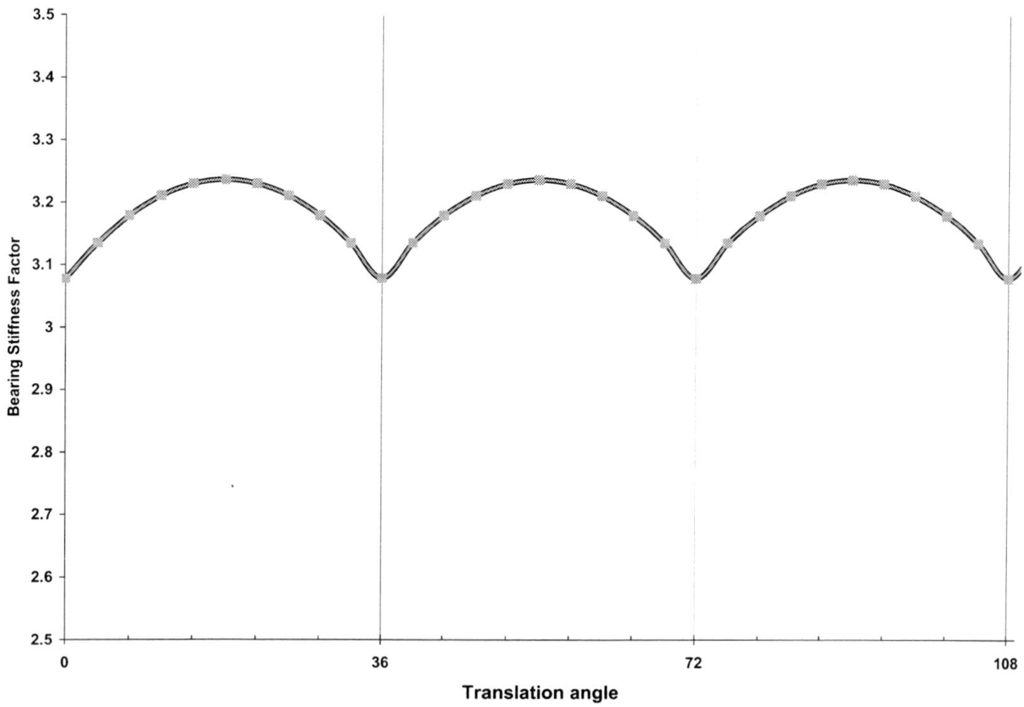

Figure 2.10. Bearing stiffness function.

Coefficients a_i are function of the number of balls under load, and ϕ represents roller translation angle.

The solution of the dynamic model requires the definition of the transmitted force *P(t)*. Ideally, it should be constant, and equal to the radial force. But, it is not the case; first of all, the radial force varies according to every application, and the rolling bearing itself produces a specific type of excitation forces. These forces are associated with physical defects on the bearing, and there are basically four types of excitation forces:

- Pitting or cracks on the rolling element (F_{re})
- Defects on the inner race (F_{ir})
- Damages on the outer race (F_{or})
- Damages on the rolling element casing (F_c)

Each force has its particular frequency, there have been published different approaches to determine them, but they can be summarized as:

- Inner race frequency ω_{ir}

$$\omega_{ir} = \frac{N}{2}\left(1 + \frac{db}{Dp}\cos(\alpha)\right)\omega \tag{2.19}$$

- Outer race frequency ω_{or}

$$\omega_{or} = \frac{N}{2}\left(1 - \frac{db}{Dp}\cos(\alpha)\right)\omega \tag{2.20}$$

- Case frequency ω_c

$$\omega_{c} = \frac{1}{2}\left(1 - \frac{db}{Dp}\cos(\alpha)\right)\omega \tag{2.21}$$

- Rolling frequency ω_{re}

$$\omega_{re} = \frac{Dp}{db}\left(1 - \left(\frac{db}{Dp}\cos(\alpha)\right)^2\right)\omega \tag{2.22}$$

Where; db is the roller element diamenter, Dp is the pitch diameter, ω is the shaft's angular velocity, and α is the contact angle for axial bearings.

Finally, the external forces can be combined as:

$$P(t) = F_{ir}\cos(\omega_{ir}) + F_{or}\cos(\omega_{or}) + F_{c}\cos(\omega_{c}) + F_{re}\cos(\omega_{re}) \tag{2.23}$$

If we modeled a simple rotor supported by two rolling bearings (Figure 2.11), the dynamic equations can be expressed as:

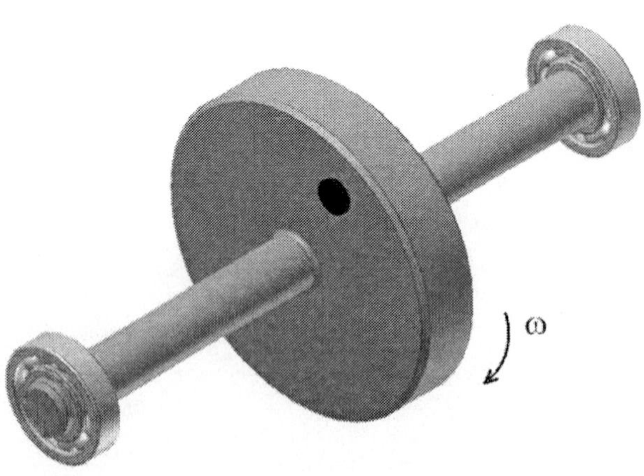

Figure 2.11. Unbalanced shaft mounted on two rolling bearings.

$$m\ddot{x} + c\dot{x} + k_x(\phi) = me\Omega^2 \cos(\Omega t) + F_{ir} \cos(\omega_{ir}) + F_{or} \cos(\omega_{or}) + F_c \cos(\omega_c) + F_{re} \cos(\omega_{re}) \tag{2.24}$$

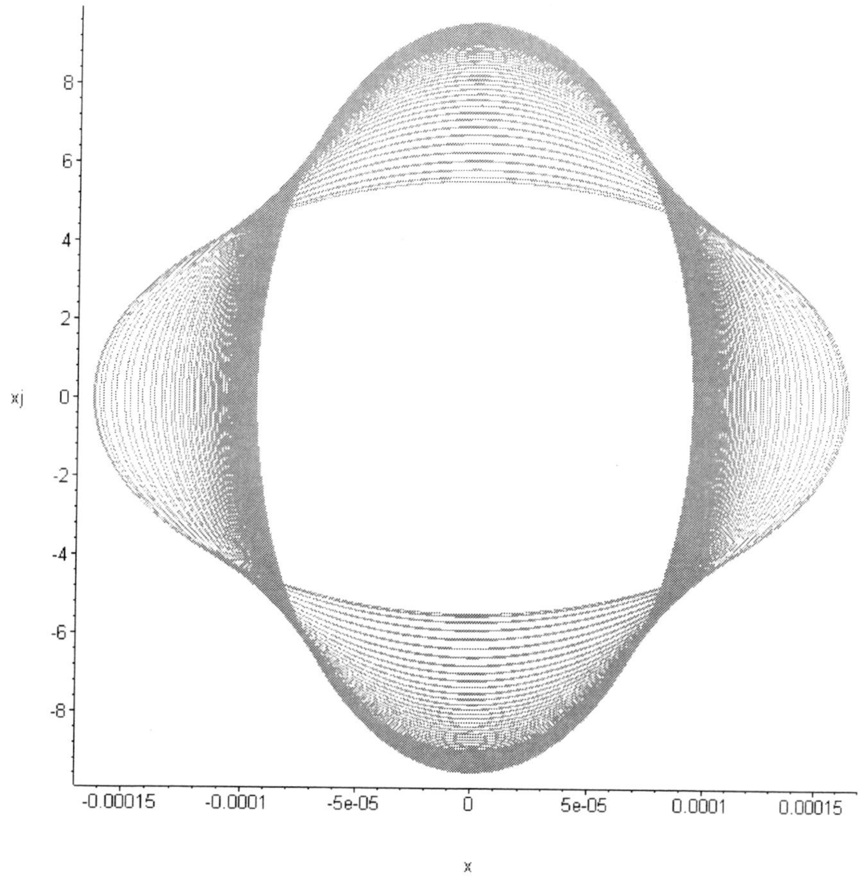

Figure 2.12. Free vibration of a simple supported shaft mounted on two ball bearings.

As discuss in the gear box example, the phase diagram is a clear indication of the nonlinear behavior of a shaft mounted on rolling bearings, it creates to attracting poles, as shown in Figure 2.12. It is also important to notice that stiffness nonlinearity can be modeled with a Duffing's equation.

Solving the nonlinear equation, it is possible to identify the effects of each of the faulty external forces. Next plots illustrate the phase diagrams of these effects. It is important to notice that the diagram changes its shape when there is a defect on the bearing.

The harmonic distortion effect is clearly identified as the solution rotates. This diagram sets the reference for the analysis of the external excitations. Each excitation modifies the shape and it is a way of identifying failures on the bearing.

It is clear form Figure 2.13 that the system shows a chaotic response when the excitation force is near the resonance frequency.

Figure 2.14 represents the effect of two defects on the phase diagram, (a) corresponds to a defect on the inner ring, and (b) to a defect on the outer ring. In both cases the rotor speed is low compare to its resonance frequency.

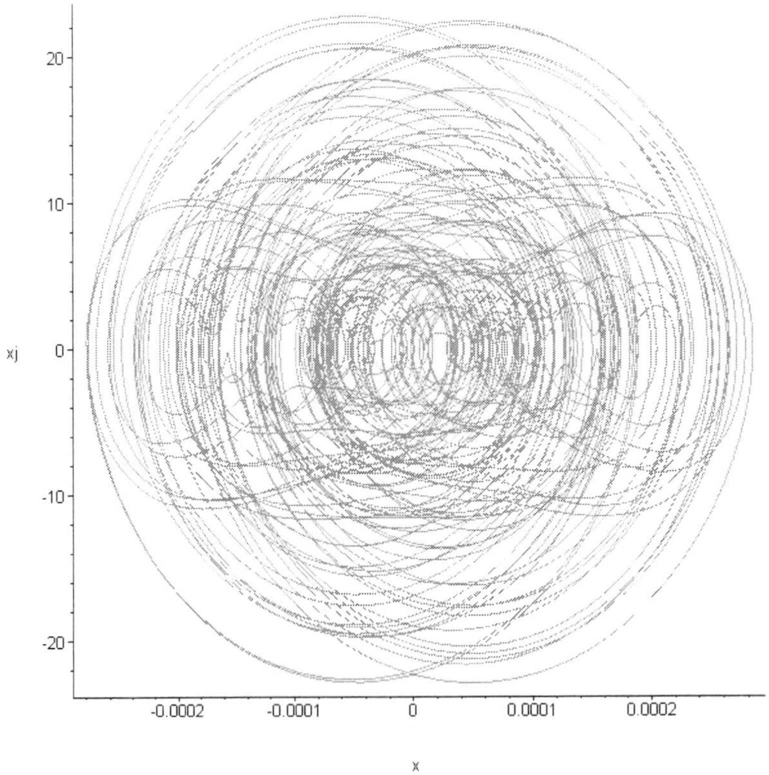

Figure 2.13. Phase diagram of an unbalance rotor near its resonance frequency.

With these results, it is clear that rolling bearings present a non-linear behavior due to the discontinuities on their stiffness; this discontinuity is the result of the translation of the rolling element around the shaft. One of the main characteristic of their dynamic response is that the stiffness is a function of the frequency instead of only function of the displacement, as in other mechanical systems. Therefore, the phase diagram presents a different pattern.

One of the challenges of a monitoring system is the identification of early faults in rolling bearings. As we have described before, failures in bearings start at surface level; thus, they generate a relatively small energy vibration compared to other sources, and its identification is very cumbersome. With the application of phase diagram plots, early failures can be predicted in real time. The process is as follows:

Vibrations are measured with a transducer, preferably an accelerometer. Then, the signal is analogically integrated in real time. Then, in an oscilloscope, the phase diagram is plotted. When the equipment is new, the first diagram corresponds to the healthy reference, (in Chapter 5, we present different "shapes" of a phase diagram). Since we know that bearings have a nonlinear response, and that this response is the result of its stiffness dependency on frequency, we can monitor the phase diagram in order to "see" the moment where instabilities occur. In this way, if we permanently monitor the "shape" of the phase diagram, and we detect the appearance of instabilities, then we will be able to detect early faults. Figure 2.16 show a phase diagram of a healthy bearing. In this figure, we can see four mayor loops, they correspond to the main frequencies, the unbalance load produces the external loop, and the other three are the mayor bearing frequencies. This diagram shows similar shapes at different time steps.

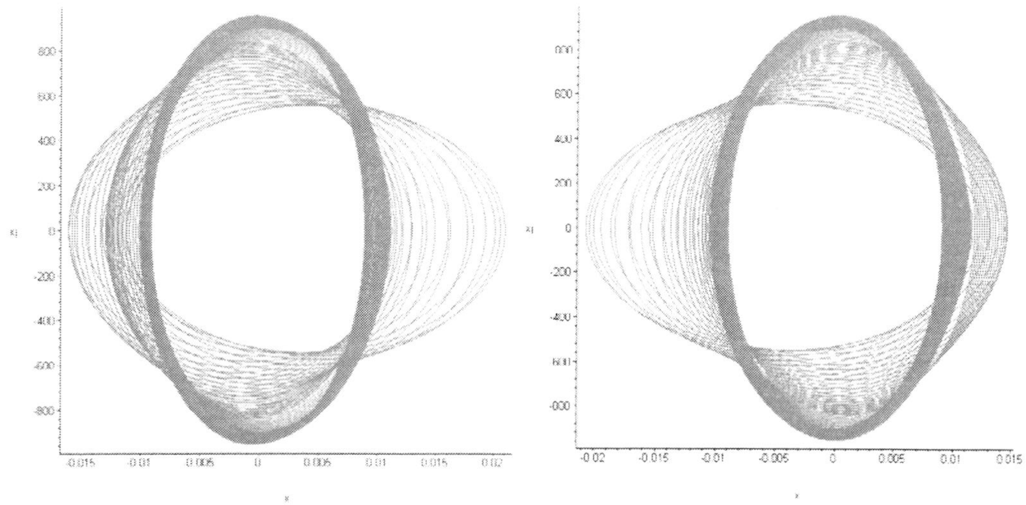

(a) (b)

Figure 2.14.

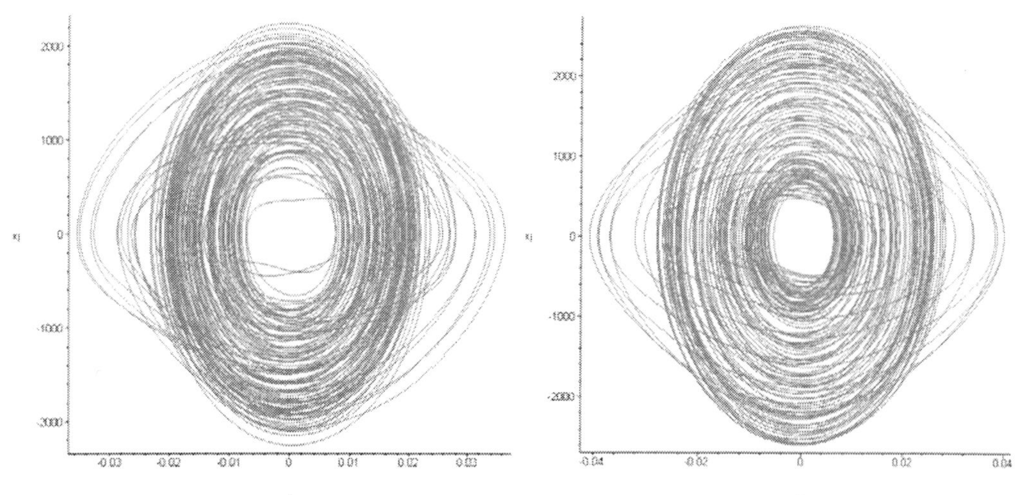

(a) (b)

Figure 2.15. Phase diagram, at high speed (1/4 resonance frequency).

Figure 2.17 shows the same phase diagram, but in this case, the bearings are damaged. As it can be seen, there is a dominant loop, which stands for one frequency, and several small loops that is the result of the nonlinear instabilities. These small loops vary in number and shape at different time steps; therefore, it can be concluded that the phase diagram can identify early bearing nonlinear stabilities earlier, and faster, than other monitoring techniques.

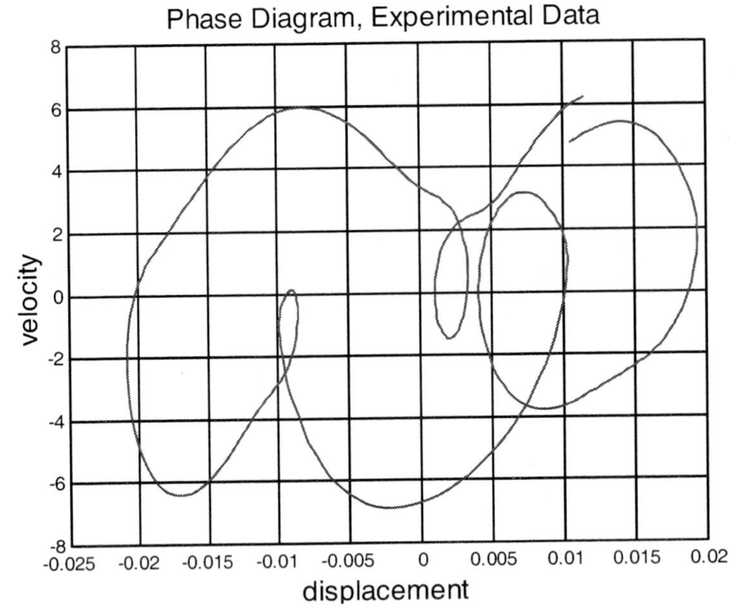

Figure 2.16. Phase diagram of a healthy bearing. Experimental data.

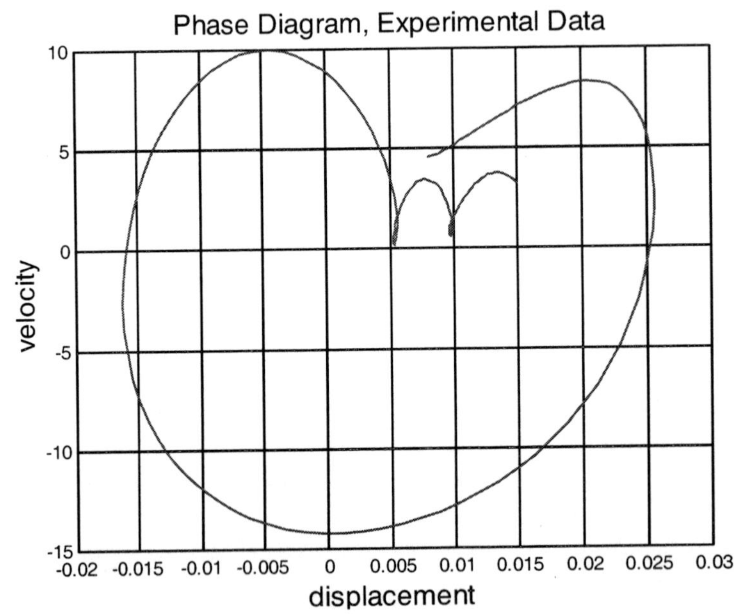

Figure 2.17. Phase diagram of a damage bearing. Experimental data.

Sliding Bearings

Sliding bearings are key elements on modern machine tools. As a result of their construction the sliding bearings have very low friction and very high accuracy. They are formed by a fixed track and traveling block that are link through a set of metallic balls that circulate within the traveling block as it slides along the track. They have a wide range of applications since they allow very fast displacements and they can be combined in different arrangements in order to have orthogonal movements. (Figure 2.18)

One of the disadvantages of the sliding bearings is their low damping value; this effect causes the existence of small displacements along the perpendicular directions of the translation axis. These small displacements produce damaging vibrations and are the result of the backlash between the rolling elements and their elastic deformation. When the system is exited with an external forced, it shows a nonlinear response. This nonlinearity is typical of a system with discontinues stiffness. There are two types of discontinuities in sliding bearings: the backlash between the roller elements and the tracks, and the circulation of the rolling elements. This circulation varies the number of elements that support the load and, thus the stiffness changes as a function of the traveling displacement.

Thus, the dynamic model of a sliding bearing is derived from the assumption that the stiffness is only due to the rolling elements deformation and it is proportional to the number of balls in contact. Therefore, since the number of rolling elements varies as the moving element travels along the track, the stiffness of the system is a function of the traveling displacement.

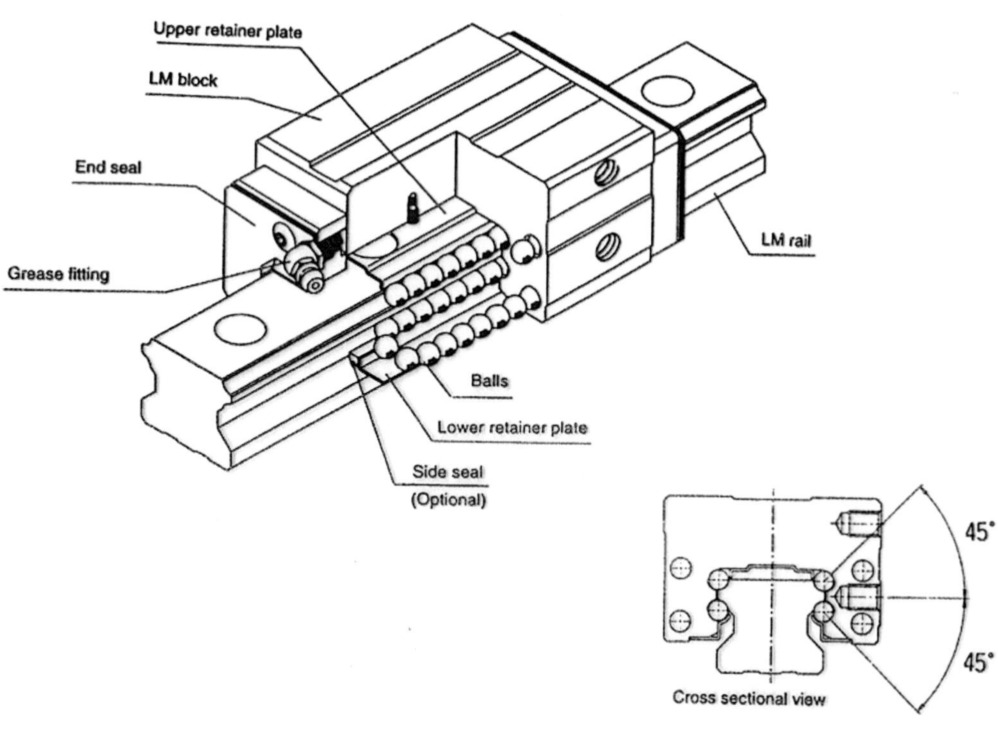

Figure 2.18. Construction of a sliding bearing.

We applied the same procedure as in rolling bearings, the stiffness of each rolling element under the external load is determined using Hertz' formula.

$$K = \frac{F_a}{\delta_a} \tag{2.25}$$

Thus, the deformation along the line of action is calculated as:

$$\delta_a = \frac{0.002}{\sin\alpha}\left(\frac{Q^2}{D}\right)^{1/3} \tag{2.26}$$

Where Q is the applied force, D is the ball's diameter and α is the angle of the track design. The applied load Q is a function of the number of rolling elements in contact and is defined as:

$$Q = \frac{F_a}{N\sin\alpha} \tag{2.27}$$

We have shown that the stiffness of the sliding bearing is a function of the displacement, thus it is non constant. This variation can be approximated using Fourier's series.

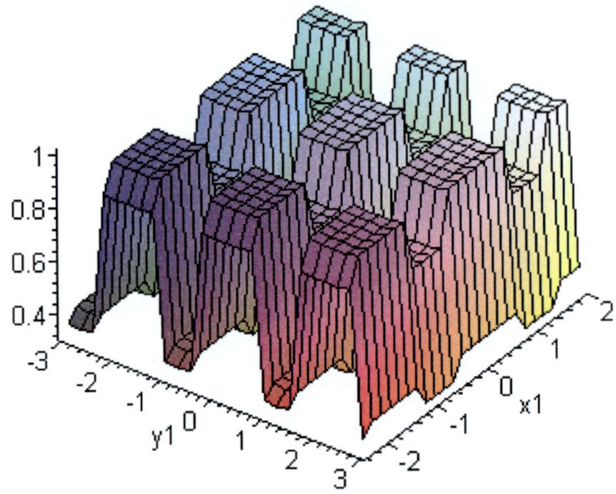

Figure 2.19. Stiffness map of a two degrees of freedom sliding system.

Therefore, in this case, the number of rolling elements in contact can be estimated as:

$$\begin{aligned} N &= \sum_{n=0}^{\infty} a_n \cos(2\pi f n x) + \sum_{n=1}^{\infty} b_n \sin(2\pi f n x) \\ &= Z_0 + Z_1\cos(x) + Z_2\sin(x) \end{aligned} \tag{2.28}$$

Since the displacements are small compared to the traveling displacement, the series can be simplified using a Taylor´s approach as:

$$N = Y_0 + Y_1\, x + Y_2\, x^2 + Y_3\, x^3 \tag{2.29}$$

This form of the stiffness is similar to those of the other systems described before, and demonstrates that a mechanical system with sliding bearings behaves as a nonlinear system. Also, they can be analyzed with nonlinear dynamic theories described in Chapter 1, as the bifurcations theory.

If we combine two sliding bearings in order to form a orthogonal table with displacements in the X-Y directions, the stiffness of the system will have a pattern similar to the one described in Figure 2.19.

If we simulate a orthogonal X-Y table as a lumped mass model, the sliding table could be represented as the system shown in Figure 2.20. The stiffness of the sliding bearings is approximated as nonlinear springs, and the mass of the table is supported along the perpendicular axes (with respect to the traveling axis).

Using the Hamilton's principle the system of equations is derived as:

$$m\begin{bmatrix}1 & 0\\ 0 & 1\end{bmatrix}\begin{Bmatrix}\ddot{y}\\ \ddot{z}\end{Bmatrix} + c\begin{bmatrix}1 & 0\\ 0 & 1\end{bmatrix}\begin{Bmatrix}\dot{y}\\ \dot{z}\end{Bmatrix} + \begin{bmatrix}K_y & K_y K_z\\ 1 & K_z\end{bmatrix}\begin{Bmatrix}y\\ z\end{Bmatrix} + \begin{Bmatrix}O(y^2)\\ O(z^2)\end{Bmatrix} + \begin{Bmatrix}O(y^3)\\ O(z^3)\end{Bmatrix} = \{F(t)\} \tag{2.30}$$

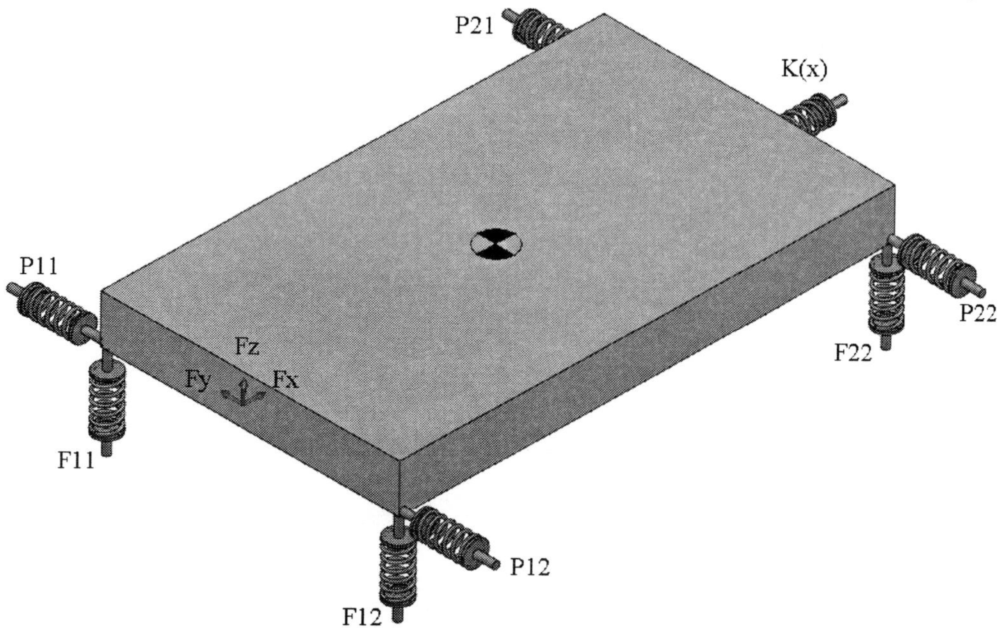

Figure 2.20. Lumped mass model of an orthogonal table.

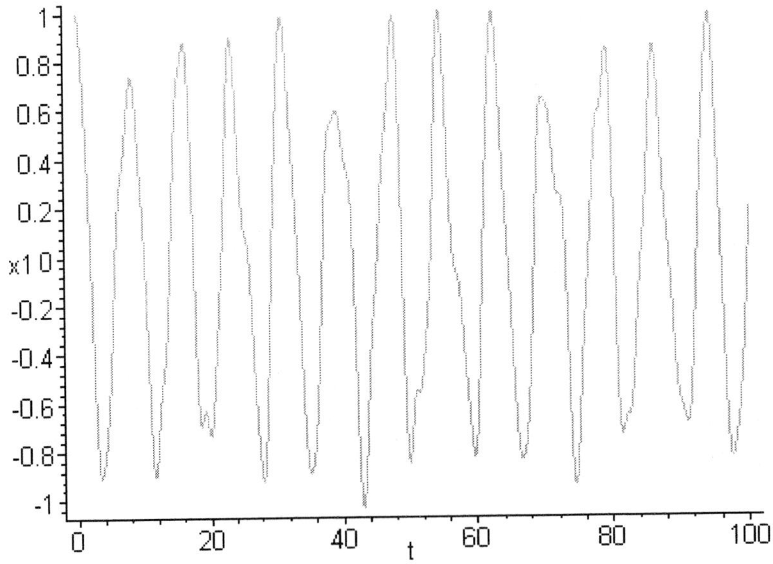

Figure 2.21. Time response in the x direction.

Where K_y, K_z are the nonlinear springs that represents the sliding bearing.

The solution of the system shows a bifurcation, similar to the gear case. Figure 2.21 illustrates the time response, where the nonlinear response is evident.

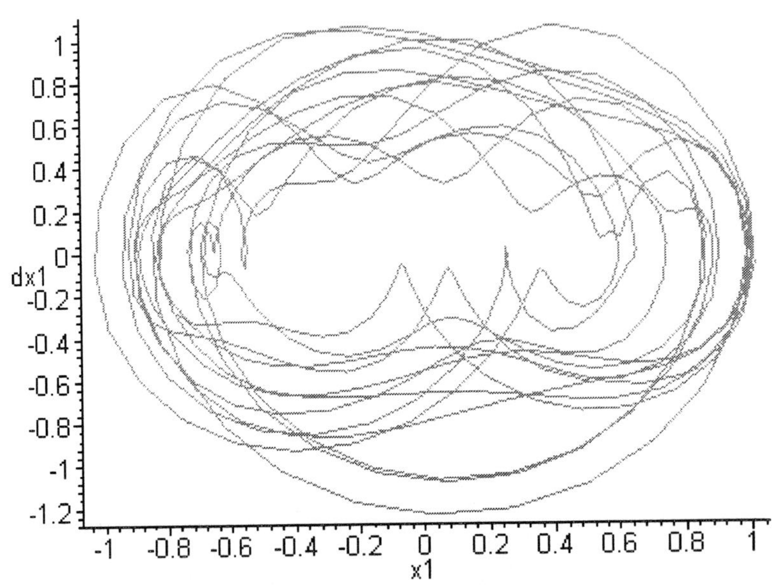

Figure 2.22. Phase diagram of the orthogonal table with initial conditions x_0=0, $\dot{x}_1 = 1$.

Analyzing the phase diagram of the system (Figure 2.20) it is clearly identified the presence of two attracting poles. This is a clear indication of the existence of a bifurcation of the solution. In practice, this situation implies that the system will have an unstable response in a very wide range of frequencies; thus, the operation conditions of the orthogonal table will be limited to a narrower range. With this model, it is easy to identify the operation conditions and the unstable conditions of any machine tool. The excitation frequencies come from the cutting force of the tool.

We discuss in Chapter 1 that the nonlinear dynamic systems depend upon the initial conditions, if we solve the same system with different initial conditions, the response is completely different as shown if Figure 2.23 and 2.24.

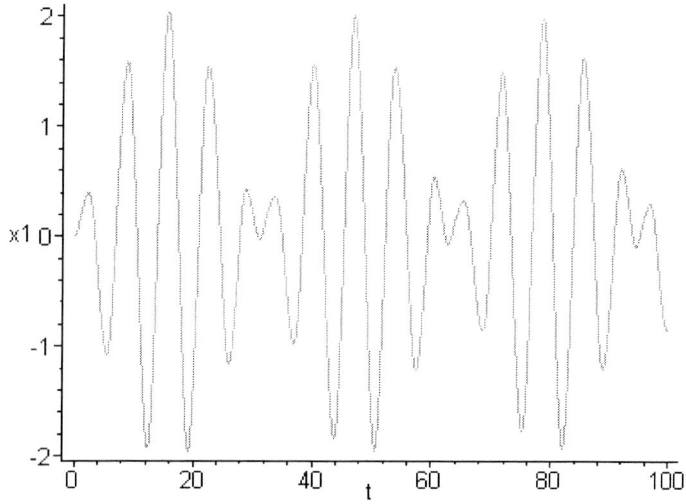

Figure 2.23. Time response of the orthogonal table with initial conditions x_1=0, $\dot{x}_1 = 0$.

The phase diagram also shows a different pattern:

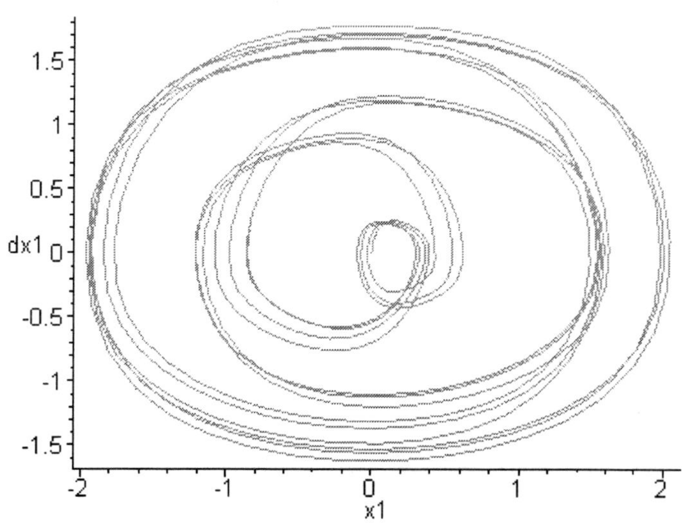

Figure 2.24. Phase diagram of the orthogonal table with initial conditions x_1=0, $\dot{x}_1 = 0$

Backlash

It is a mechanical problem that can be seen also as a discontinuous problem. When the driven force has fluctuations, backlash plays a fundamental role in the system's dynamics since the resistance load advances over the driven force and instantly it reverses the direction of motion, at this instant, there is no contact between the mechanical elements, and it can be considered as a discontinuity on the system's stiffness. Although there is no direct relation between backlash and nonlinear effects as bifurcation or chaos, one of the reasons can be that stiffness discontinuities is directly associated to these phenomena.

Backlash is a problem particular to each machine. Even two apparently identical machines can have complete different behaviors due to differences in their backlash. It is also associated to impacts. When a mechanical system moves in alternating directions, there is a loose of contact. Backlash is a time dependant problem, rather than a position dependant problem. Except for certain elements, backlash function is difficult to established, it depends on the operating conditions and machine cycles, a way to represent it is as a transient function of the form

$$x_{nl}(t) = \sum_{n=1}^{N} a_n \delta_\tau e^{-b_n(t-\tau)} \cos(\omega_n (t-\tau) + \phi_n) \tag{2.31}$$

In Chapter 5 we will discuss the application of time-frequency maps to identify this type of signals.

ROTOR RUBBING

There are other dynamic effects that present stiffness discontinuities as a result of a defect. The most significant one is rotor rubbing. This phenomenon is undesirable, and it is considered as a machine failure. Although it may not make a machine to stop, it will create other problems that eventually will be very dramatic. Rubbing has different patterns; they depend on the force of contact between the rotor and the static part it is hitting. At low contact forces, rubbing generates waves that propagate through the structure. Although originating from a different mechanical process, this wave motion is similar to the wave energy that propagates from microscopic cracks within solid structures.

A number of reasons can be identified for the onset on light rubbing in a rotating machine. These include thermal effects, foundation movement, component movement, rotor unbalance or misalignment. First, partial rubbing consists on intermittent contact occurring instantaneously within the period of the shaft rotation. Secondly continuous rubbing involves more sustained contact between shaft and surrounding components. Although the mechanisms by which rub phenomena escalate are complex, it is suggested that for machinery of higher rotational speeds, a high concentration of partial rub events can lead to more sustained rubbing, which in turn can induce more serious vibration. Light rubbing is difficult to identify and will not be discussed in this book. Here we will deal with more severe rubbing as can be found in large machines such as turbines or large gear boxes.

From the analytical point of view, most of the mathematical models are based on the Jeffcott rotor model. The impact force takes into consideration Coulomb friction and neglects its damping effect. The impact between the rotor and the stator occurs during a short period of time (meaning that it rubs only during a fraction of the circumferential movement).

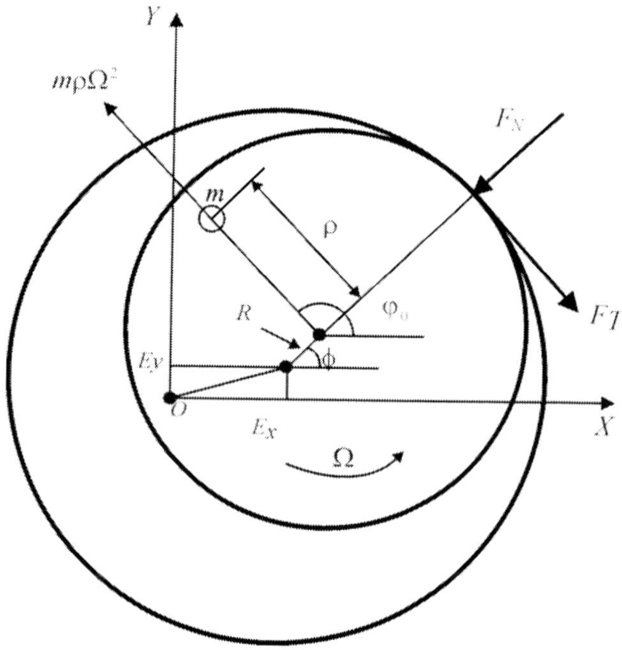

Figure 2.25. Rotor-rubbing model.

From Figure 2.25, normal and tangential forces are calculated as:

$$F_N = \begin{cases} 0, & R \le c \\ K_s(R-c), & R \ge c \end{cases}$$
$$F_T = \mu F_N \tag{2.32}$$

Representing these forces into a Cartesian system,

$$F_x = -F_N \cos\phi + F_T \sin\phi$$
$$F_y = -F_N \sin\phi - F_T \cos\phi \tag{2.33}$$

Therefore, the motion equations for the rotor system can be expressed as:

$$M\ddot{x} + C\dot{x} + Kx = F_x + m\rho\Omega^2 \cos(\varphi_0 + \Omega t)$$
$$M\ddot{y} + C\dot{y} + Ky = F_y + m\rho\Omega^2 \sin(\varphi_0 + \Omega t) \tag{2.34}$$

Since the normal force F_N is non-continuous, the system stiffness varies with respect to rotor angular position. Thus, the model presents a non-linear behavior that can be related to the stiffness variation. This variation can be approximated as a rectangular periodic function (Figure 2.26). As in previous cases, the stiffness coefficient can be represented as a cubic polynomial.

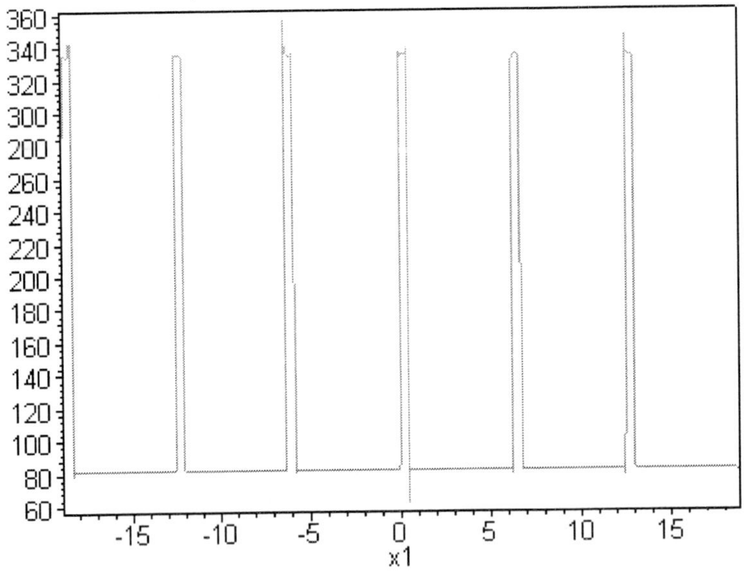

Figure 2.26. Stifness variations as a function of the angular displacement.

From a practical point, it is very difficult to determine the instant when the rotor contacts a fixed surface (casing or bearing) and the time span of this contact. Therefore, the non-linear coefficients can not be estimated on real time basis, and the non-linear model can only be used for analysis, not for monitoring.

Before contact, the rotor stiffness is calculated from shaft stiffness, during contact, the stiffness of the system increments, the increment can be calculated using Hertz' contact stress. Thus, adding the contact stress, the system stiffness varies as a function of the position.

It has been reported in the literature analyses of steady-state vibration for identifying rotor rubbing. In these reports the analysis is made out the frequency spectrum, if a rotor operates below its critical speed, the spectrum will show frequencies at half its rotating frequency with several harmonics, if the rotor operates above its critical speed, sub-harmonics at 1/3 of the rotating speed will appear. But FFT is unable to identify the instant when the rotor hits a fixed surface. These sub-harmonic frequencies are the result of the numerical treatment of the nonlinear time response.

Therefore, the need of better analysis tools brought the application of other techniques such as wavelets (see Chapter 5). They have been applied to nonlinear vibration and transient analysis because they combine time and frequency analysis. They can localize a time-frequency window along any signal and, in this way, the instant of hit. In the case of a rotor rubbing the casing or the bearing, wavelets allow the identification of the "wave shape" through a predetermined time-frequency function and the instant when rubbing occurs. The challenge of this method is the definition of the function that represents the actual phenomenon.

The system of equations is solved numerically or using the Average Value Method. With the numerical solution, the displacement and velocity are calculated as a function of time. With these values, we construct the phase diagram (Figure 2.27)

Analyzing the phase diagram of a rotor rubbing the casing, it is clear the presence of two attracting poles, as discuss in chapter 1; the presence of attracting poles in the phase diagram is a clear indication of the existence of bifurcations.

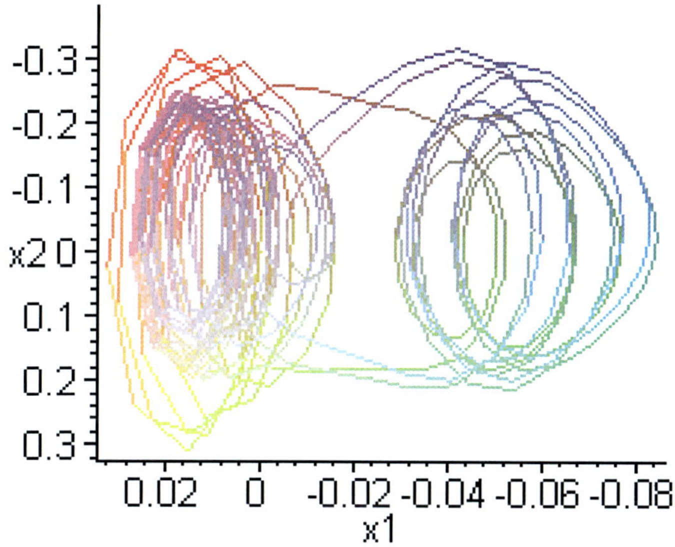

Figure 2.27. Phase diagram of a rotor rubbing the casing.

The solution of the system with the mean value method allows us the calculation of the displacement amplitude as a function of the frequency. In Chapter 1, the mean value method was introduced. It was demonstrated that the amplitude function can be expressed as:

$$\Omega = \frac{-3\alpha r^3 \pm 4\sqrt{\gamma^2 - (\omega\delta r)^2}}{4r} = \frac{\omega_0}{\varepsilon}\left(1 - \frac{\omega}{\omega_0}\right) \tag{2.35}$$

In Figure 2.28 the amplitude as a function of the frequency is plotted. Analyzing the solution, it is easy to identify that rotor rubbing presents the jump phenomenon.

From these two analyses, it is clear that rubbing is strongly nonlinear. Its effects could be catastrophic, especially in high speed machines as aircraft engines. Numerical analyses help understanding the phenomenon, with them it is possible to set operating conditions that will prevent the presence of rubbing, and they will indicate the frequencies of the dynamic response, but in a practical application, it is quite difficult to identify it. In the case of rubbing, at early stages, the dominant signal is the transient vibration but when presents a permanent response, the dominant signal is the nonlinear response. Based on these types of responses, time-frequency and wavelet analysis are the most appropriate technique for analyzing measurement data.

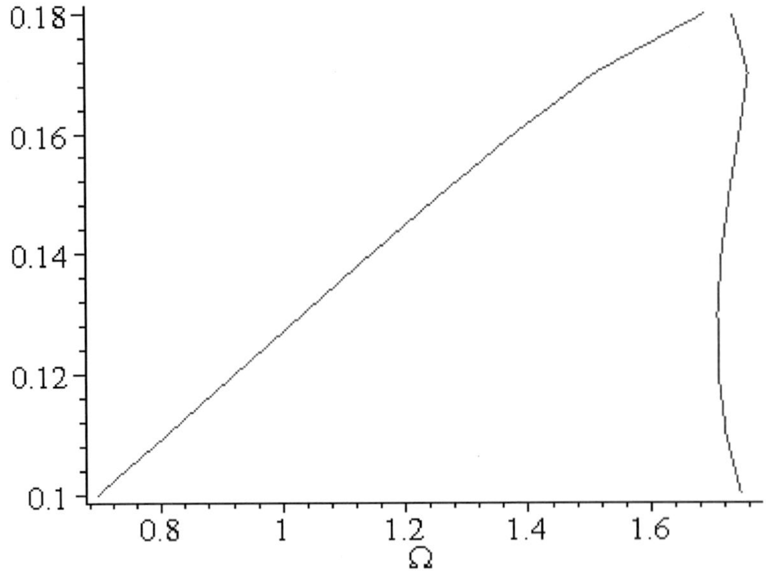

Figure 2.28. Rotor rubbing also presents the jump phenomenon.

To show the difference between an analysis based on FFT and a time-frequency analysis, we include an example of measurements from a rotor rubbing the casing. In this case, we measured the rotor acceleration in a radial direction after it has developed a "strong" contact with the casing. It was reported that when the force created by the rub impact is weak, sub-harmonics and super-harmonics appear. When the impact is relatively strong, only super-harmonics are dominant. For detecting these effects from measured data, we used the Wigner-Ville Wavelet Distribution. The Wigner-Ville Distribution introduces cross-terms when the signal contains multiple frequencies. However, this effect can be avoided by filtering the signal and using the Hilbert transformed version. In this way, frequency spectrums are separated as a function of time. Then a time-frequency map is constructed, as shown in Figure 2.29.

Results show that low energy components and unsteady signals can be highlighted. But both methods have the limitations that time variation has to be induced running the machine through a ramp-up test, and it is not always practical to do so. In this way, the jump phenomenon is reproduced, and different amplitude levels are clearly identified on the time-frequency map. It is important to notice that rubbing excites high frequencies, that differ from any harmonic; but also, analyzing the amplitude as the speed ramps up and down, it can be distinguished that the amplitude value its lower at the ramp down.

The method can be extrapolated to any rotating machinery. First, the well known frequencies must be identified (namely the designed frequencies). In almost any rotating system this can be done using the design information. Then the vibration signal is analyzed with the wavelet transform.

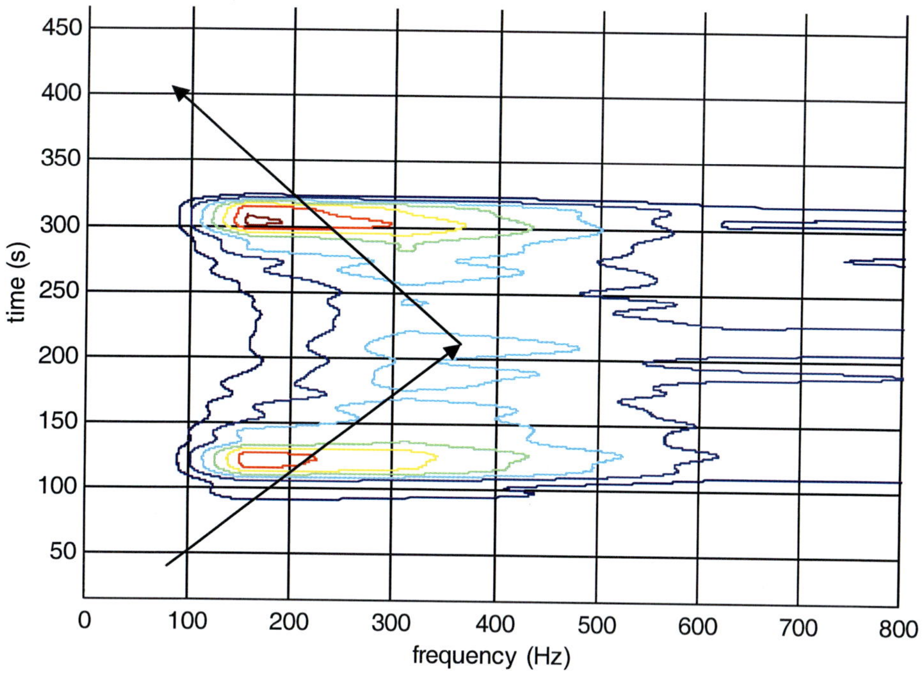

Figure 2.29. Time-frequency map of a run-up and run-down test.

The two arrows on Figure 2.29 indicate the speed increments for the ramp up, and the speed reduction for the ramp down. It can be notice that around 175 Hz the highest amplitude values appear. The important aspect of this graph is that they have different amplitude; therefore, we can conclude that there is a nonlinear effect similar to a jump phenomenon.

From practical data, it is the easiest way to identify jump phenomena without running different tests at different conditions.

Chapter 3

STIFFNESS DISCONTINUITIES DUE TO LARGE DEFORMATIONS

INTRODUCTION

It is difficult to identify continues systems that present nonlinear stiffness. In this chapter we describe systems as trust elements of slender beam structures. The dynamic equations are derived using the basic principles of the theory of elasticity and Galerkin's approximation. It is found that these structures can also be modeled with a Duphing's type equation, thus, they behave in a similar manner as the systems described in Chapter 2. Slender structures are been employed in many mechanical systems, they are present in high speed gas turbines blades, wind mills blades and other structures.

In the first part, we develop the nonlinear formulation for a trust element. With this example we illustrate the nonlinear behavior of a simple trust, and it could be extended to any solid element. The form of the equation and its solution leads us to the conclusion that only beam and plates elements could model slender elements. To get to this formulation the full radius of rotation must be included into the strain field. The second part of this chapter deals with analysis of slender beams using nonlinear beam elements. We include different examples in order to illustrate the application of the method, and we analyze to cases: a simple degree of freedom beam, and a multiple degree of freedom system.

TRUST ELEMENTS

In this section we applied large deformations in order to determine the non-linear dynamic model of solid elements. We found that this consideration generates a mathematical model that has minimum applications in dynamic problems. The non-linear terms have significant effect only with extremely large deformations. Thus, it implies that the material will go into plastic deformation. In the case of ultra elastic materials some other theories, like viscoelasticity behaviors, may present a better approximation to practical applications.

The development of the model for solid elements is included for fulfillment of this work. Its formulation is different from beams and plates elements, because the strain theory has the same degrees of freedom as the geometrical coordinates. In this case the orthogonal displacements are parallel to spatial coordinates, thus, for a simple trust element, is displacement u occurs along the coordinate x.

A general form of the strain field as a function of the generalized displacements with large deformations is:

$$\varepsilon_{ij} = \frac{1}{2}\left(\frac{\partial u_i}{\partial x_j} + \frac{\partial u_j}{\partial x_i}\right) + \frac{1}{2}\left(\frac{\partial u_j}{\partial x_i}\right)\left(\frac{\partial u_i}{\partial x_j}\right) \tag{3.1}$$

where i=x,y,z and j=x,y,z.

For a trust element with a single degree of freedom $i=j=x$, $x_i=x_j=x$, and $u_i=u_j=u$. Therefore:

$$\varepsilon_{xx} = \left(\frac{\partial u}{\partial x}\right) + \frac{1}{2}\left(\frac{\partial u}{\partial x}\right)^2 \tag{3.2}$$

Applying a finite element approach, the field displacement can be approximated by a linear polynomial,

$$u(x) = \sum_{i=1}^{n} a_i x^{i-1} \tag{3.3}$$

Considering that the trust element can be modeled as a linear element with two nodes per element, $n=2$

Then, the polynomial coefficients are:

$$\begin{aligned} a_1 &= \frac{1}{l}\left(u_1 x_2 - u_2 x_1\right) \\ a_2 &= \frac{1}{l}\left(u_2 - u_1\right) \end{aligned} \tag{3.4}$$

where u_1 represents the displacement of the first node, and u_2 represents the displacement of the other node.

The equations of motion are derived from the Hamilton's principle. Although Hamilton´s principle and Lagrange´s equation are equivalent, Hamilton´s principle is broader in the sense that it applies to distributed systems. In this case, it is assumed that it holds for an individual element as

$$\frac{d}{dt}\left(\frac{\partial L^e}{\partial u_{ti}}\right) - \frac{\partial L^e}{\partial u_i} = 0 \tag{3.5}$$

where

$$L^e = T^e - W^e \tag{3.6}$$

and

$$u_t = \frac{du}{dt} \tag{3.7}$$

It can be shown that the kinetic energy of each element can be expressed as:

$$T^e = \frac{1}{2}\int_V \rho u_t^2 dV \tag{3.8}$$

The other terms in Hamilton's principle combine two terms: The required work to deform the element (internal energy), and the work done by the external forces.

$$W^e = W_E^e - W_I^e \tag{3.9}$$

The deformation energy can be calculated as:

$$W_I^e = \frac{1}{2}\int_V \sigma_{ij}\varepsilon_{ij} dV \tag{3.10}$$

where sum on repeated subscripts is implied

In this chapter we only consider elements with linear material properties, thus the constitutive equation can be expressed as:

$$\sigma_{ij} = E_{ij}\varepsilon_{ij} \tag{3.11}$$

And, for the trust element

$$\sigma_{xx} = E\varepsilon_{xx} \tag{3.12}$$

Including eq. (12) into eq. (10), the deformation energy for the trust element becomes:

$$W_I^e = \frac{1}{2}\int_V E\varepsilon_{xx}^2 dV \tag{3.13}$$

on the other hand, the work done by the external forces depends upon an axially distributed pressure f and the generalized nodal loads:

$$W_E^e = \int fudx + \sum_1^m Q_i^e u_i dx \tag{3.14}$$

Finally, we substitute eq. (3.8), (3.13) and (3.14) into eq. (3.5), and we separate terms that only depend on u_{ti}, from those that depend on u_i, thus, Hamilton's principle is written as:

$$\frac{d}{dt}\left(\frac{\partial T^e}{\partial u_{ti}}\right) + \frac{\partial W^e}{\partial u_i} = 0 \tag{3.15}$$

Developing each term, for the kinetic energy we found that:

$$\frac{d}{dt}\left(\frac{\partial T^e}{\partial u_{ti}}\right) = \frac{d}{dt}\left(\frac{\partial T^e}{\partial u_t}\frac{\partial u_t}{\partial u_{ti}}\right) = \frac{d}{dt}\int_V \rho u_t \left(\frac{\partial u_t}{\partial u_{ti}}\right) dV \tag{3.16}$$

Assuming that mass can be concentrated at each node:

$$\frac{d}{dt}\left(\frac{\partial T^e}{\partial u_{ti}}\right) = m_i u_{tti} \tag{3.17}$$

where m_i is the corresponding node mass and

$$u_{tti} = \frac{d^2 u_i}{dt^2} \tag{3.18}$$

Developing the second term from equation (3.5):

$$\frac{\partial W^e}{\partial u_i} = \left(\frac{\partial W^e}{\partial a_i}\right)\left(\frac{\partial a_i}{\partial u_i}\right) \tag{3.19}$$

For the trust element:

$$\frac{\partial W_I^e}{\partial a_2} = \frac{EAl}{2}\left(2a_2 + 3a_2^2 + a_2^3\right) \tag{3.20}$$

and

$$\frac{\partial a_2}{\partial u_1} = -\frac{1}{l}$$
$$\frac{\partial a_2}{\partial u_2} = \frac{1}{l} \tag{3.21}$$

Finally substituting eqs. (3.20) and (3.21) into (3.19), and rearranging every term, the equations of motion are

$$m_1 u_{tt1} + \frac{EA}{2}\left[2\left(\frac{u_2 - u_1}{l}\right) + 3\left(\frac{u_2 - u_1}{l}\right)^2 + \left(\frac{u_2 - u_1}{l}\right)^3\right] = \int f dx + Q_1$$
$$m_2 u_{tt2} - \frac{EA}{2}\left[2\left(\frac{u_2 - u_1}{l}\right) + 3\left(\frac{u_2 - u_1}{l}\right)^2 + \left(\frac{u_2 - u_1}{l}\right)^3\right] = \int f dx + Q_2 \tag{3.22}$$

In the case of elements with more than two nodes, the number of equations is the same as the number of nodes).

It can be proven that for solid elements the individual system of equations can be written as:

$$[m]\{u_{tt}\} + [K]\{u\} + \{f(u^2)\} + \{g(u^3)\} = \{Q\} \tag{3.23}$$

The nonlinear terms appear only when the displacement is larger than half the trust length. Thus, they can not be applied to slender elements because the non-linear terms have significant effect only with very large deformations. In such cases, these deformations could not exist in practical applications, only systems constructed with materials like rubber could present this type of deformations. For those cases, a better approach will be the use of viscoelastic models, where the stiffness is related to the elastic terms, and damping is related to the viscous terms (See Chapter 4).

BEAM ELEMENTS

Due to their basic formulation beam elements represent slender structure in a simplified way. They are considered as one-dimensional element, but they combine both in line displacements (as trust elements) and perpendicular displacements due to its bending. Since we have discussed the effects of in line displacements (previous section), we will present only the nonlinear bending terms. It is shown that the nonlinear terms come from the bending curvature instead of the Taylor's series expansion of trust elements. In this sense, the strain function derives from its basic definition of bending curvature; thus, we fully expand curvature terms to reach the nonlinearities.

The strain field as a function of the bending curvature is expressed as:

$$\varepsilon_{xx} = z\frac{\partial^2 w}{\partial x^2}\left(1+\left(\frac{\partial w}{\partial x}\right)^2\right)^{-3/2} \tag{3.24}$$

where z is the local coordinate perpendicular to the beam dimension, x is the coordinate along the beam dimension and w is the displacement along z.

Expanding the rational term as a convergent series

$$(1+b)^{-3/2} = 1-\frac{3}{2}b+\frac{15}{8}b^2-\frac{105}{48}b^3+\ldots \tag{3.25}$$

Then, the strain function can be rearranged as:

$$\varepsilon_{xx} = z\frac{\partial^2 w}{\partial x^2}\left(1-\frac{3}{2}\left(\frac{\partial w}{\partial x}\right)^2\right) \tag{3.26}$$

Similarly as previous section, the field displacement can be approximated by a linear polynomial:

$$w(x) = \sum_{i=1}^{n} a_i x^{i-1} \tag{3.27}$$

Considering a one-dimensional beam element with two nodes, there are four generalized coordinates, namely:

$$\mathbf{w} = \{w_1, \theta_1, w_2, \theta_2\} \tag{3.28}$$

where w is the displacement of nodes 1 and 2, and θ is bending angle at both nodes.

Therefore, the polynomial has four coefficients. Substituting equation (3.27) into equation (3.26), the strain field is expressed as:

$$\varepsilon_{xx} = z(2a_2+6a_3x)\left(1-\frac{3}{2}(a_1+2a_2x+3a_3x^2)^2\right) \tag{3.29}$$

The equation of motion is found by applying Hamilton´s principle. In this case, the inner work is:

$$W_I^e = \frac{1}{2}\int_V \sigma_{xx}\varepsilon_{xx}dV \tag{3.30}$$

We only consider elements with linear material properties, thus the constitutive equation can be expressed as:

$$\sigma_{ij} = E_{ij}\varepsilon_{ij} \tag{3.31}$$

Since the beam element is one-dimensional

$$\sigma_{xx} = E\varepsilon_{xx} \tag{3.32}$$

Therefore, the deformation energy becomes:

$$W_I^e = \frac{1}{2}\int_V E\varepsilon_{xx}^2 dV \tag{3.33}$$

We separate terms that only depend on the generalized velocities $=\{w_{t1}, \theta_{t1}, w_{t2}, \theta_{t2}\}$, from those that depend on the generalized displacements, thus, Hamilton's principle is written as:

$$\frac{d}{dt}\left(\frac{\partial T^e}{\partial w_{ti}}\right) + \frac{\partial W^e}{\partial w_i} = 0 \tag{3.34}$$

where w_{ti} and w_i are velocity and displacement vector components.

It can be shown that the kinetic energy of each element can be expressed as:

$$T^e = \frac{1}{2}\int_V \rho w_t^2 dV \tag{3.35}$$

for the kinetic energy we found that:

$$\frac{d}{dt}\left(\frac{\partial T^e}{\partial w_{ti}}\right) = \frac{d}{dt}\left(\frac{\partial T^e}{\partial w_t}\frac{\partial w_t}{\partial w_{ti}}\right) = \frac{d}{dt}\int_V \rho w_t\left(\frac{\partial w_t}{\partial w_{ti}}\right)dV \tag{3.36}$$

Approximating the mass matrix as a diagonal matrix:

$$\frac{d}{dt}\left(\frac{\partial T^e}{\partial w_{ti}}\right) = m_i w_{tti} \tag{3.37}$$

where w_{tti} is the second time derivative of the generalized displacements.

The term associated with the potential energy can be written as:

$$\frac{\partial W^e}{\partial w_i} = \sum_j \frac{\partial W^e}{\partial a_j} \frac{\partial a_j}{\partial w_i} \tag{3.38}$$

For a beam with two nodes per element

$$\frac{\partial W_I^e}{\partial w_i} = \frac{EI}{2} \int_0^l \left(\frac{\partial}{\partial a_1} (\varepsilon z)^2 \frac{\partial a_1}{\partial w_i} + \frac{\partial}{\partial a_2} (\varepsilon z)^2 \frac{\partial a_2}{\partial w_i} \right) dx \tag{3.39}$$

Considering that the beam element is fixed at one end:

$$w_1 = 0$$
$$\theta_1 = 0$$

Thus, equation (3.39) expands into two equations:

$$K_3 = \frac{17496}{385}\frac{EIw^5}{l^7} - \frac{486}{11}\frac{EI\theta w^4}{l^6} + \left(\frac{1620}{77}\frac{EI\theta^2}{l^5} - \frac{1296}{35}\frac{EI}{l^5} \right) w^3$$
$$- \left(\frac{729}{154}\frac{EI\theta^3}{l^4} - \frac{162}{7}\frac{EI\theta}{l^4} \right) w^2 + \left(\frac{54}{77}\frac{EI\theta^4}{l^3} - \frac{216}{35}\frac{EI\theta^2}{l^3} + \frac{12EI}{l^3} \right) w$$
$$- \frac{522}{385}\frac{EI\theta^5}{l^2} + \frac{24}{7}\frac{EI\theta^3}{l^2} - \frac{6EI\theta}{l^2} \tag{3.40}$$

and

$$K_4 = -\frac{486}{55}\frac{EIw^5}{l^6} - \frac{810}{77}\frac{EI\theta w^4}{l^5} + \left(-\frac{729}{154}\frac{EI\theta^2}{l^4} + \frac{54}{7}\frac{EI}{l^4} \right) w^3$$
$$+ \left(\frac{108}{77}\frac{EI\theta^3}{l^3} - \frac{216}{35}\frac{EI\theta}{l^3} \right) w^2 + \left(-\frac{522}{77}\frac{EI\theta^4}{l^2} + \frac{72}{7}\frac{EI\theta^2}{l^2} - \frac{6EI}{l^2} \right) w$$
$$+ \frac{1944}{385}\frac{EI\theta^5}{l} - \frac{256}{35}\frac{EI\theta^3}{l} + \frac{4EI\theta}{l} \tag{3.41}$$

Including these equations into the Hamilton's principle, the equations of motion have the following form

$$[m]\{w_{tt}\} + [K]\{w\} + \{f_1(w^2)\} + \{f_2(w^3)\} + \{f_3(w^4)\} + \{f_4(w^5)\} = \{Q\} \tag{3.42}$$

For a two node element, Eq. (3.42) can be rearrange as

$$mw_{tt} + K_3 = 0$$
$$K_4 = 0 \tag{3.43}$$

where K_3 and K_4 include Eqs. (3.40) and (3.41) respectively.

The nonlinear terms reflect a hardening stiffness that includes the cubic and fifth order terms. As explained before, we include the solution of a single degree of freedom model, and a multiple degree of freedom case.

First Approach

Neglecting *θ's* higher order terms, Eq. (3.41) leads to the following simplification:

$$\theta_2 = \frac{3w_2}{2l} \tag{3.44}$$

Thus, substituting Eq. (3.44) into Eq. (3.40) the equation of motion can be expressed as:

$$mw_{tt} + EI\left(\frac{1485w^5}{385l^7} - \frac{162w^3}{35l^5} + \frac{3w}{l^3}\right) = f(t) \tag{3.45}$$

Where w represents the beam's tip displacement. This equation shows a Duffin's type equation with fifth power terms.

Equation (3.45) is solved with the Runge-Kutta's method. Figure 3.1 shows the phase diagram considering only linear terms. In this case, there is no external excitation and the initial displacement is 10 times the thickness of the beam. It is clear that the solution is stable and it has an elliptical shape. It is important to notice that the velocity and displacement scales are normalized.

Figure 3.2 shows the free vibration solution including the nonlinear terms. In this case, it is clear that the phase diagram is no longer an ellipse, but it has an almost rectangular shape. From this figure it is easy to distinguish the effects of the nonlinear terms.

Now, if we analyze the time response, it is clear that the displacement of the tip of the beam has a harmonic shape. But this response is not necessarily sinusoidal. Figure 3.3 shows the displacement function.

Analyzing the velocity and acceleration responses, it is clear that the beam behavior resembles a triangular function. The velocity shows a rectangular shape, whereas the acceleration is close to an impulse function. Figures 3.4 and 3.5 shows the velocity and the acceleration respectively.

One of the important aspects of this figure is that there is a characteristic frequency; this frequency is identified as the nonlinear frequency. Also, it can be notice that this frequency differs from the natural frequency. After analyzing the beam's free vibrations with and without nonlinear coefficients, the force vibration is studied. Two cases were analyzed; the first one corresponds to the application of an exciting force with a frequency near the linear natural frequency, for this case, Figure 3.6 shows the phase diagram. It is clear how the amplitude increases exponentially, showing resonance. For the purpose of comparison, nonlinear terms are neglected.

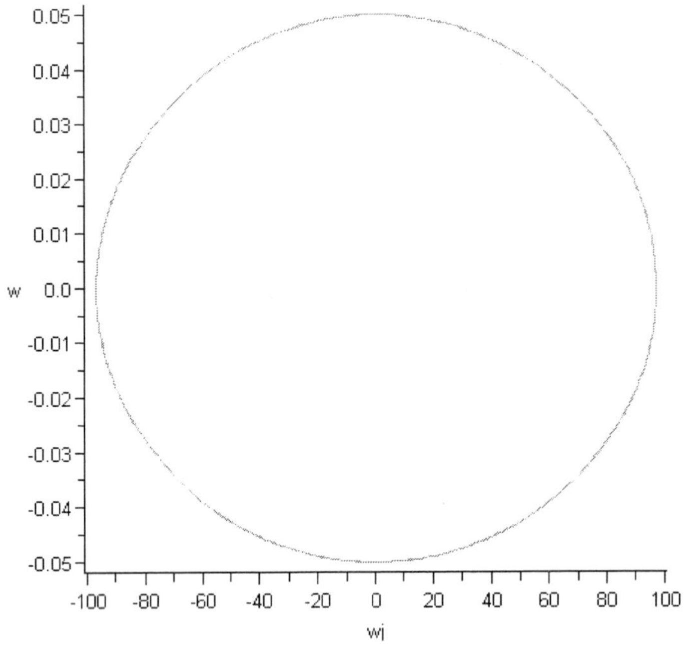

Figure 3.1. Phase diagram, linear solution.

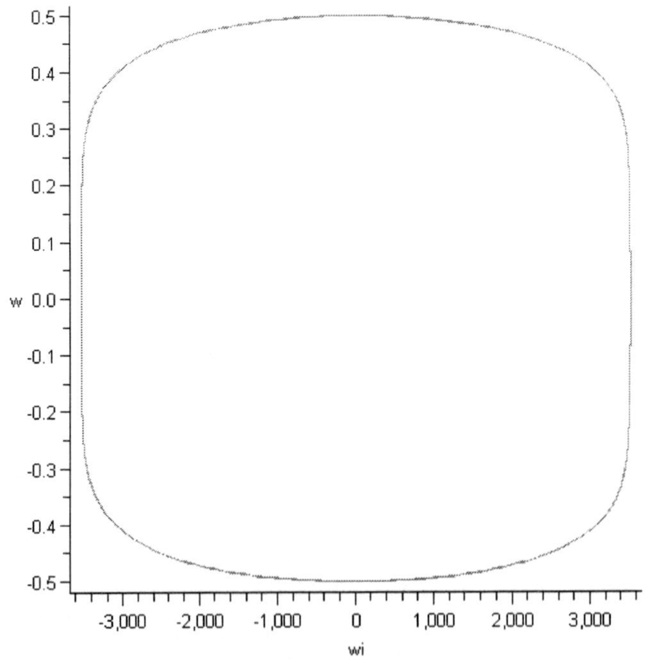

Figure 3.2. Phase diagram, nonlinear solution.

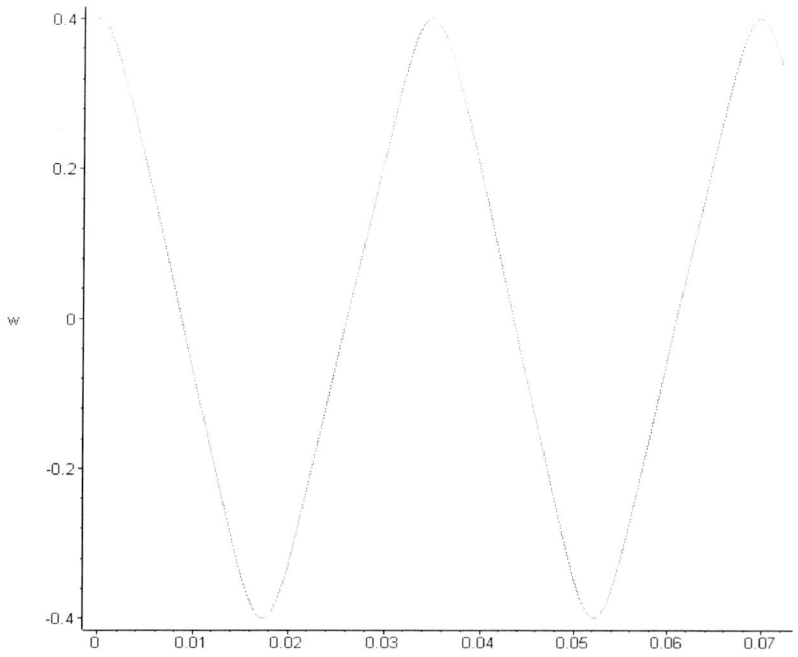

Figure 3.3. Beam displacement.

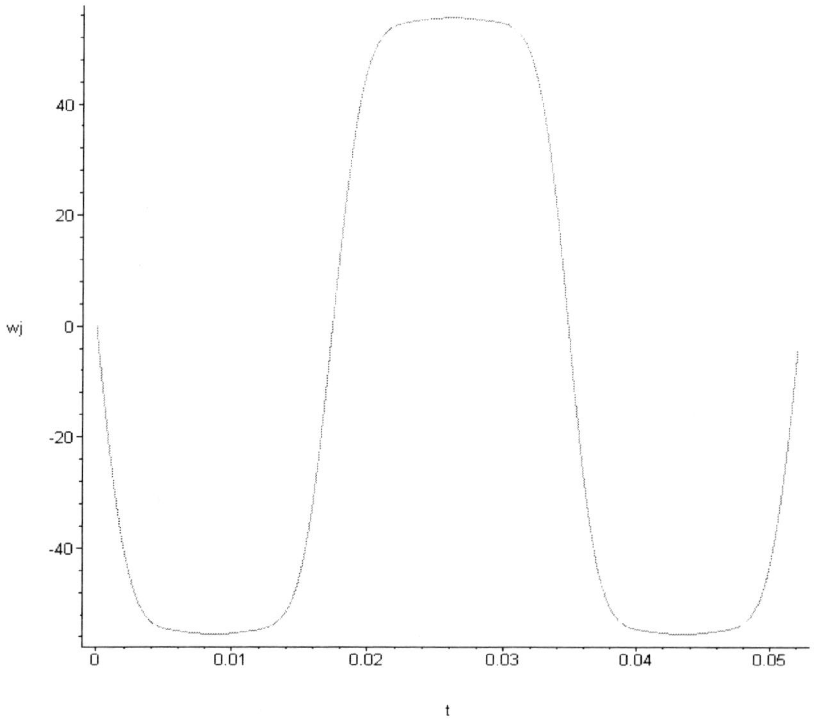

Figure 3.4. Beam velocity.

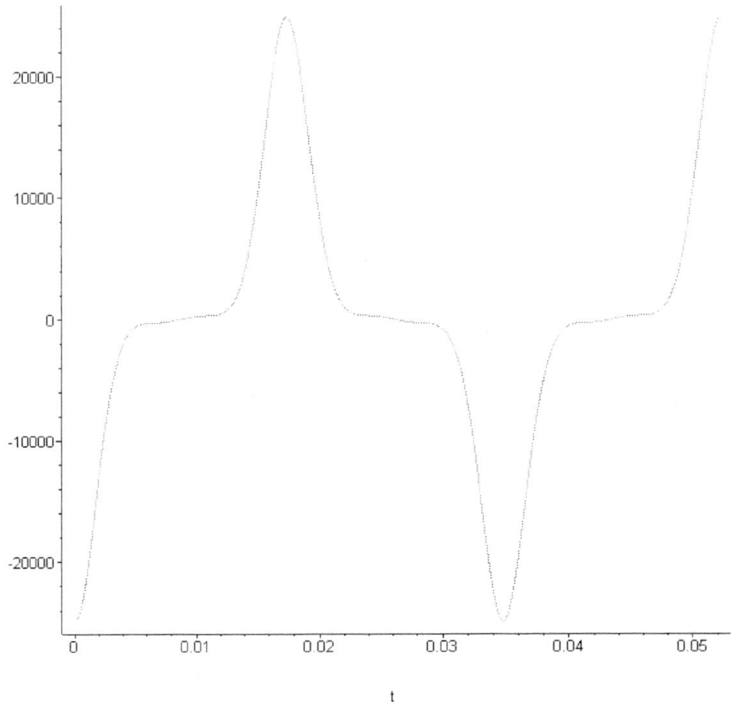

Figure 3.5. Beam acceleration.

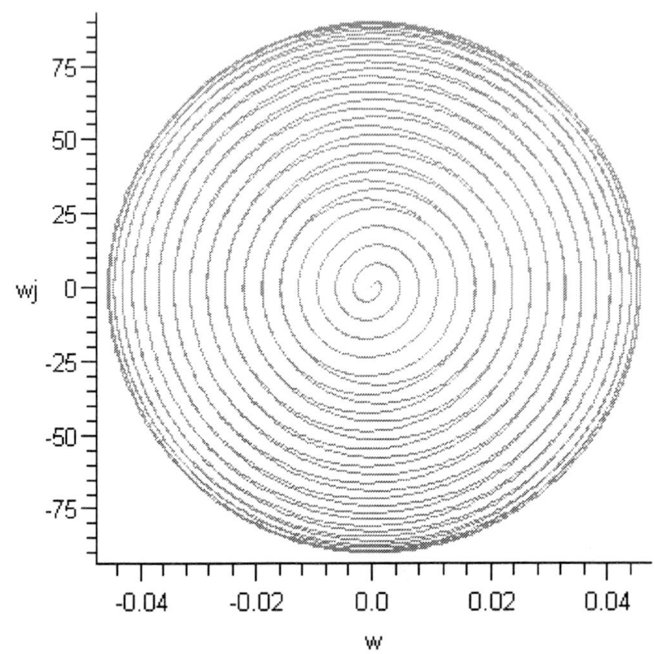

Figure 3.6. Phase diagram, linear system excitation frequency at ω_n.

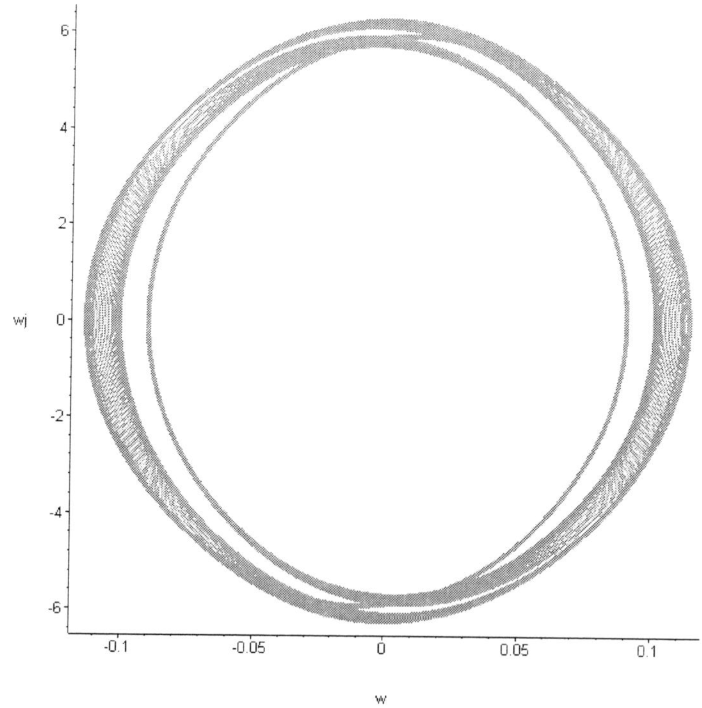

Figure 3.7. Phase diagram, excitation frequency below natural frequency.

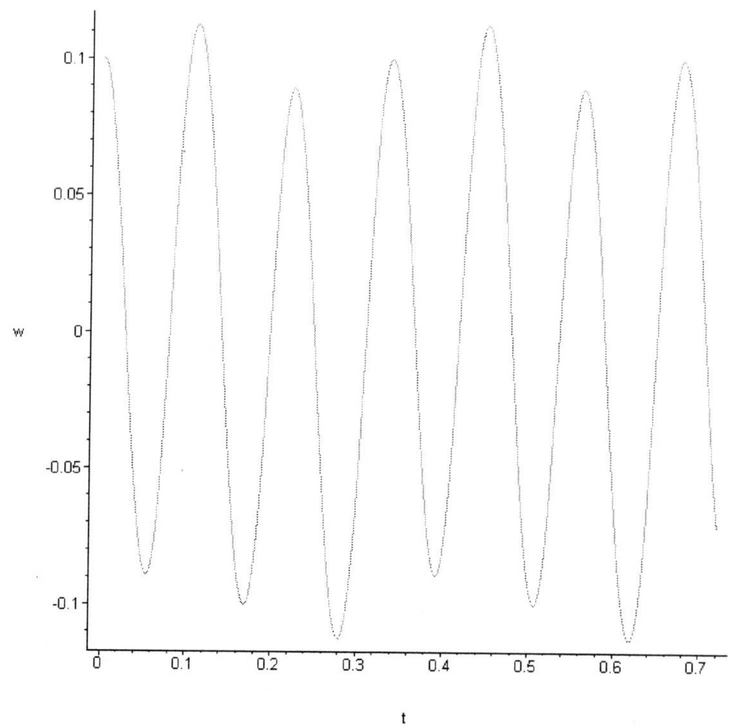

Figure 3.8. Displacement, 10% of the nonlinear frequency.

Following is the solution of the nonlinear system; a sinusoidal force is applied at different frequencies. The first application is at 10% of the nonlinear frequency. With this excitation, the system shows a slightly deform phase diagram, as shown in Figure 3.7.

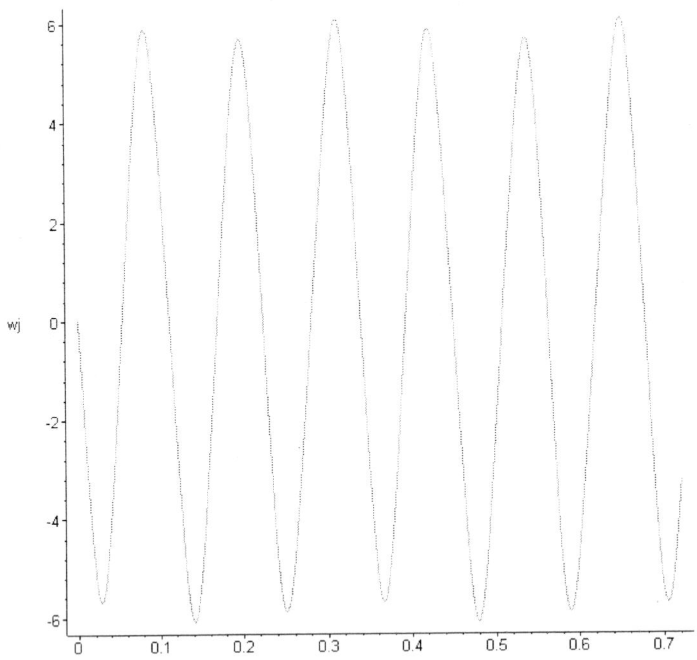

Figure 3.9. Velocity, 10% of the nonlinear frequency.

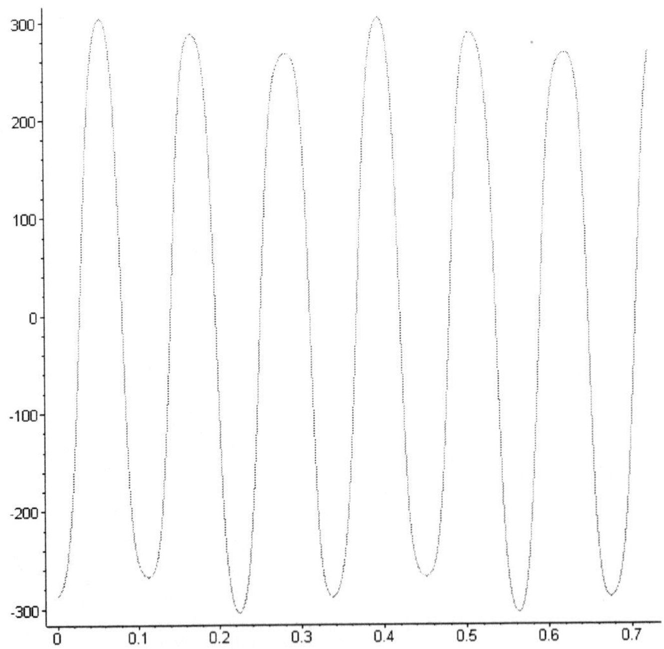

Figure 3.10. Acceleration, 10% of the nonlinear frequency.

In this representation it can be observed that the system stabilized and both frequencies are present, natural and excitation frequencies. Figure 3.8 shows the displacement of the tip. Figures 3.9 and 3.10 show its corresponding velocity and acceleration. Since nonlinear effects are not dominant, the system is then excited at the natural frequency.

In order to understand the response of the nonlinear system, we analyzed a second case; we applied a frequency near the linear natural frequency.

$$\omega_0 = 64.65 rad/s$$

Thus, the phase diagram shows the following behavior.

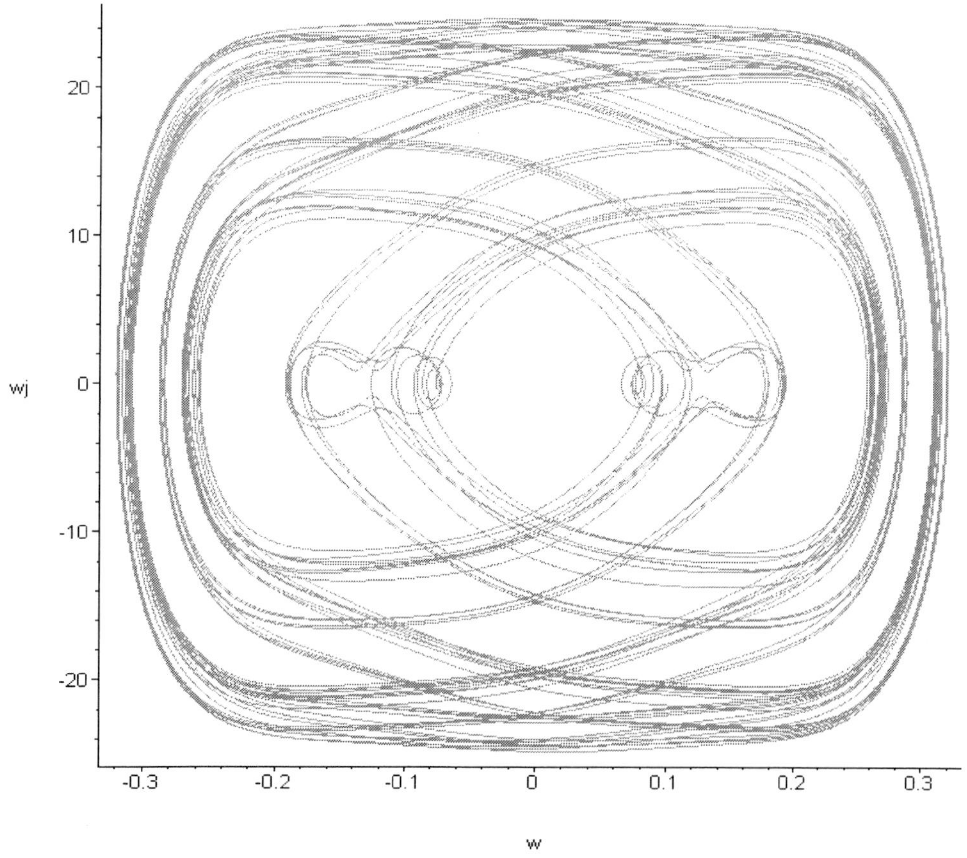

Figure 3.11. Phase diagram, force vibration with nonlinear coefficients near natural frequency.

This diagram shows the existence of two attracting poles, and the square shape caused by the nonlinear coefficients. It can be considered as a characteristic firm of a slender cantilever beam. Another important aspect of this diagram is that the sinusoidal excitation diminished and the triangular shape dominates. Particularly, if one review the shape of the acceleration response. Figures 3.12, 3.13 and 3.14 show displacement, velocity and acceleration responses as a function of time.

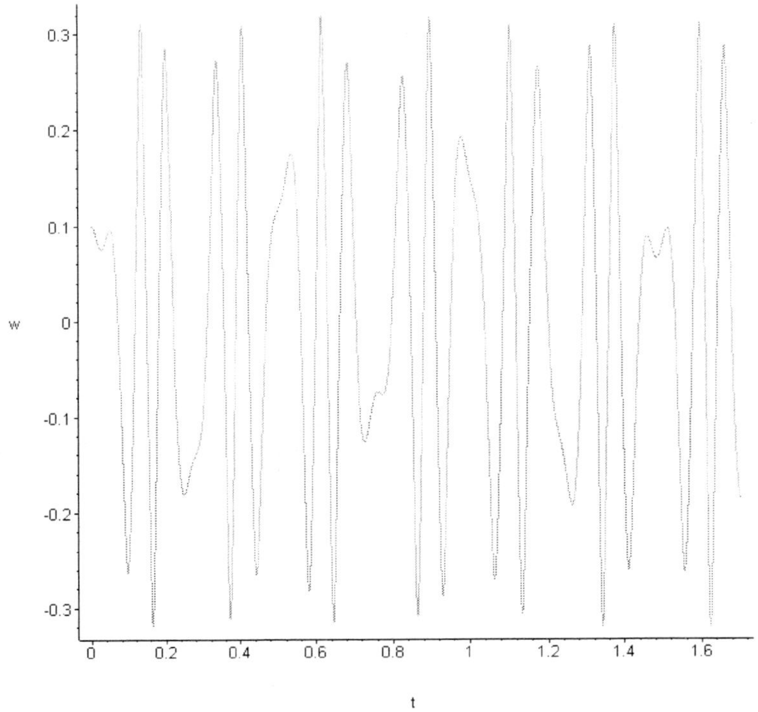

Figure 3.12. Displacement behavior of a nonlinear beam under force vibration.

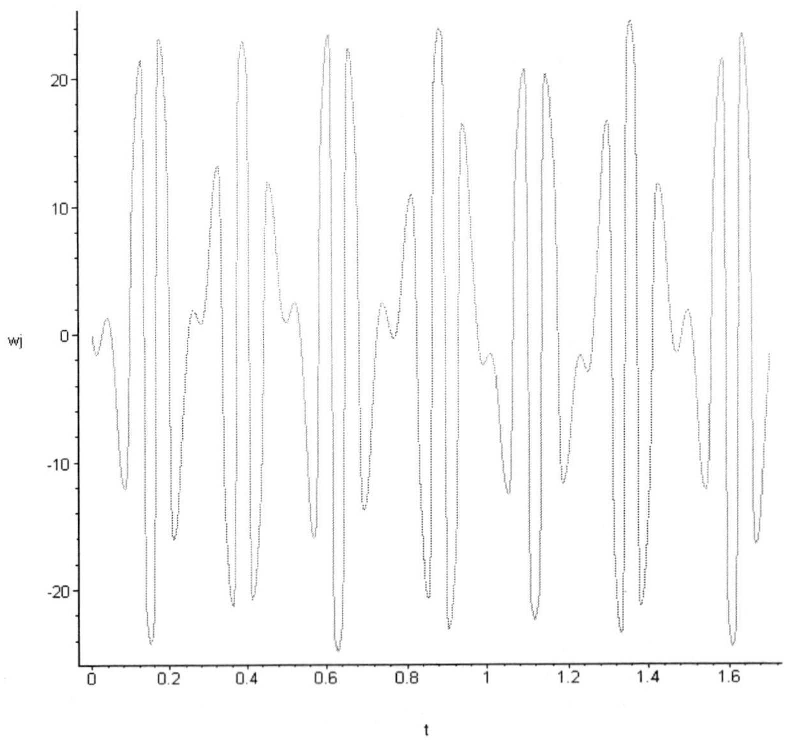

Figure 3.13. Velocity behavior of a nonlinear beam under force vibration.

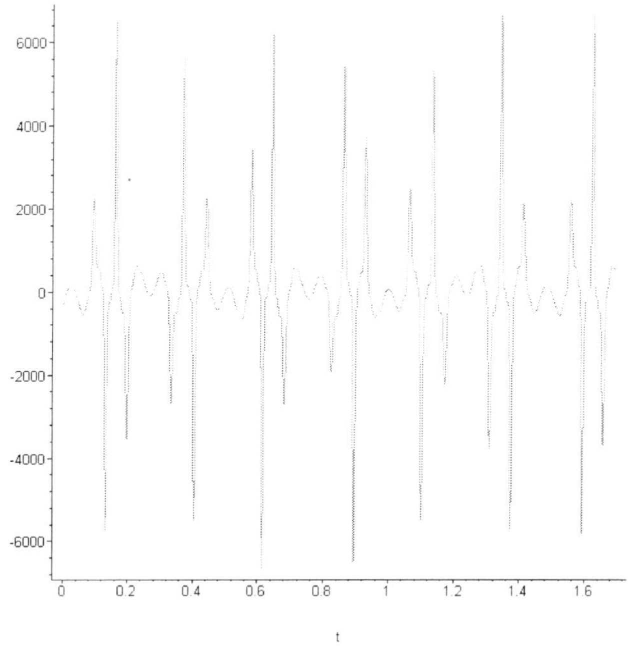

Figure 3.14. Acceleration behavior of a nonlinear beam under force vibration.

In Figure 3.14 the impulse response of the nonlinear behavior is dominant versus the external force. It is clear how the nonlinear impulse response is modulated by a second wave with a frequency close to the linear natural frequency.

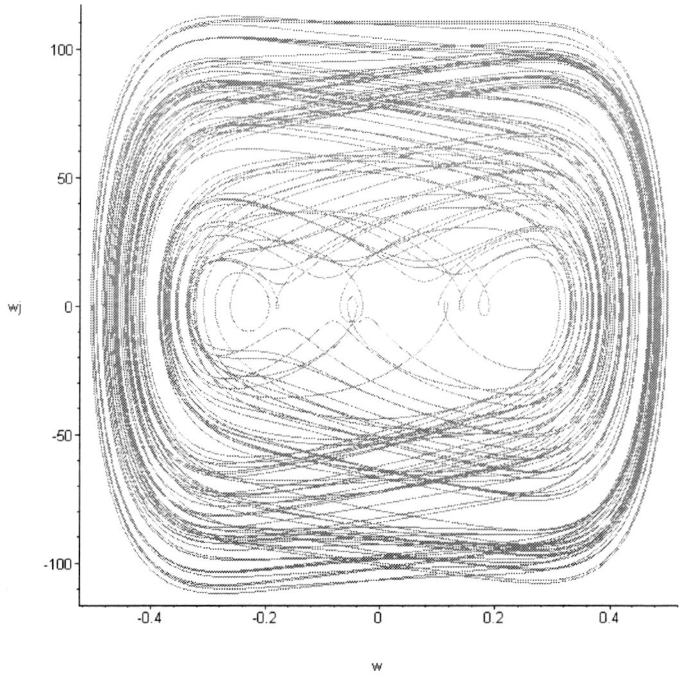

Figure 3.15. Phase diagram, near the nonlinear frequency.

The third analysis corresponds to an exciting force with a frequency near the nonlinear free response.

$$\omega_0 = 180.55 rad/s$$

In this case, the nonlinear effect dominates over the linear terms. It is important to notice that the system shows no resonance; the reason is that nonlinear terms also add a damping effect to the structure.

Once we analyzed a single degree of freedom cantilever beam, we developed the same solution for a two degree of freedom case.

Second Approach

The first approach gives a one degree of freedom solution that can be used for beams with uniform cross sections. It demonstrates the origin of the nonlinear terms and it shows that slender beams can be analyzed using Duffing's equation. With this approach, it is evident that the nonlinear behavior of a slender beam is only related to the large deformation terms, and these terms came out considering the full radius of curvature. In order to validate previous simplifications, we solved the complete system described as:

$$\begin{aligned} m\ddot{w} + K_3 &= f(t) \\ J\ddot{\theta} + K_4 &= \tau(t) \end{aligned} \tag{3.46}$$

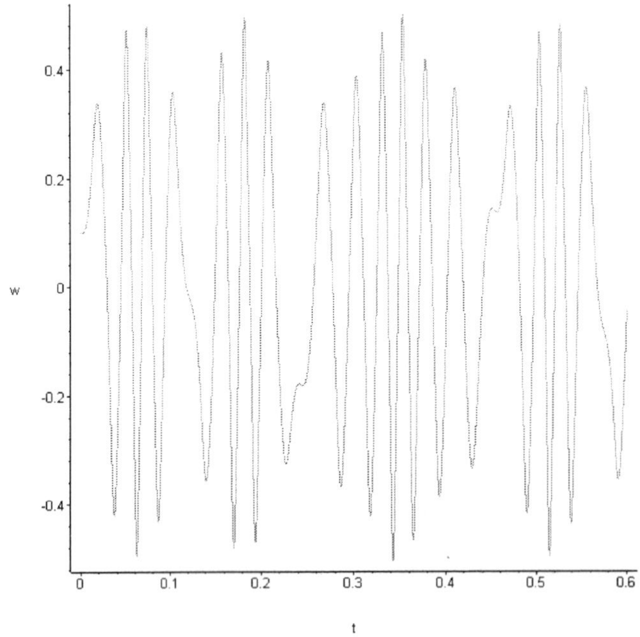

Figure 3.16. Displacement, near the nonlinear frequency.

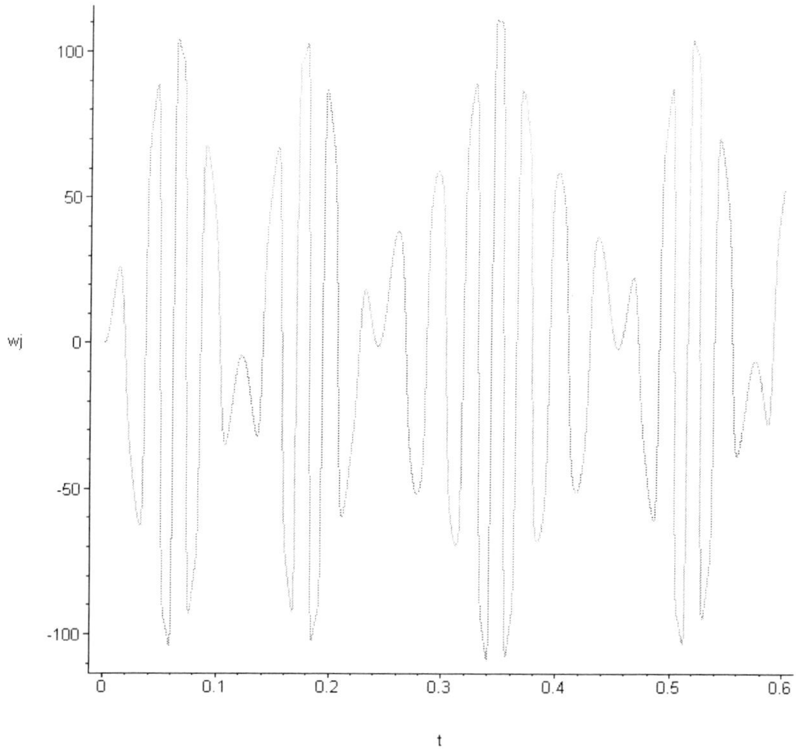

Figure 3.17. Velocity, near the nonlinear frequency.

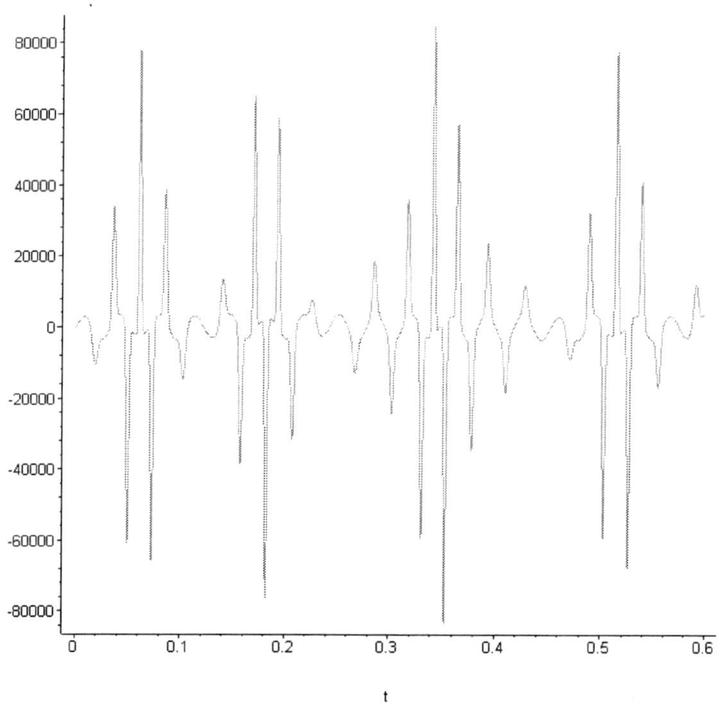

Figure 3.18. Acceleration, near the nonlinear frequency.

Where K_3 and K_4 are the terms of Eqs. (3.40) and (3.41).

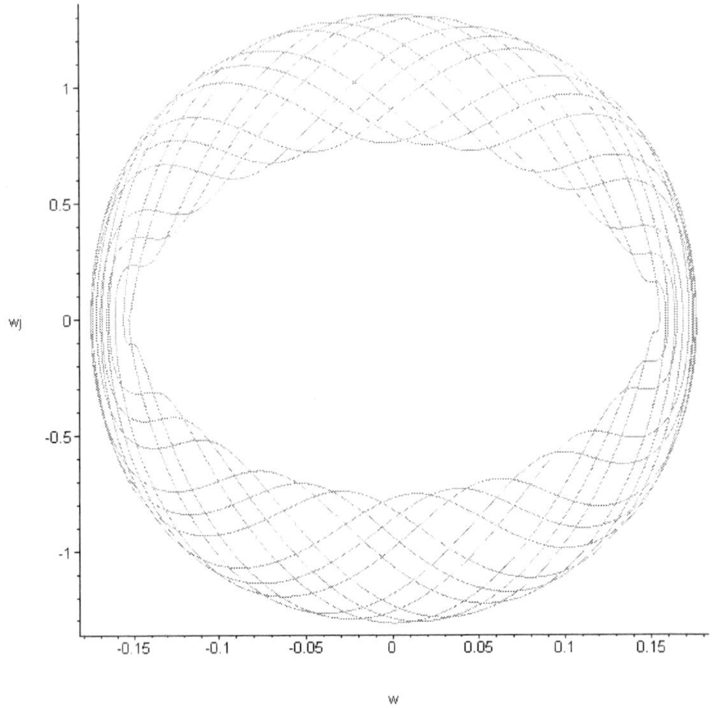

Figure 3.19. Phase diagram, two DOF free vibration linear solution.

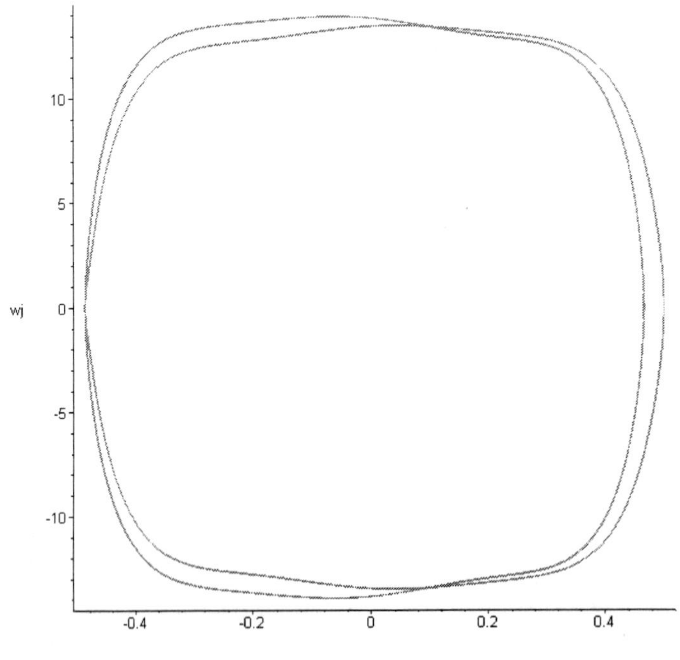

Figure 3.20. Phase diagram, two DOF nonlinear free vibration.

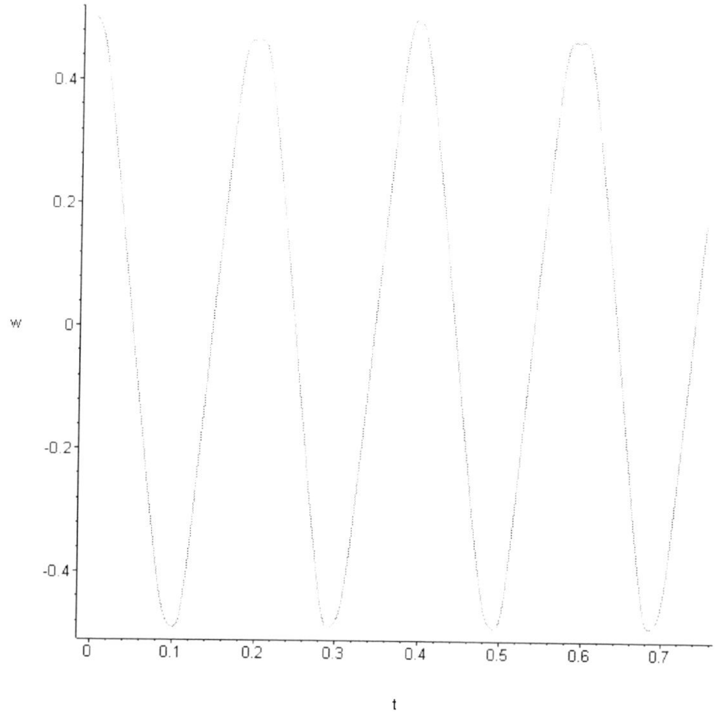

Figure 3.21. Displacement function, two DOF nonlinear free vibration.

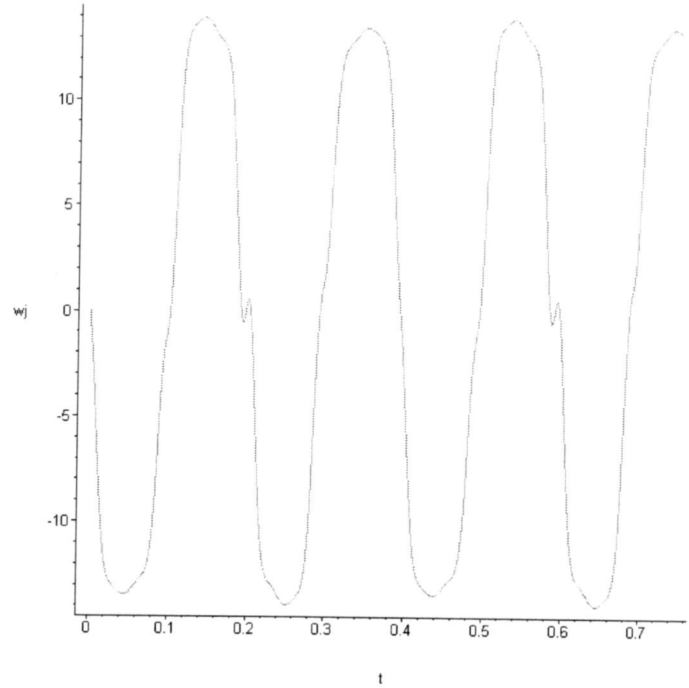

Figure 3.22. Velocity function, two DOF nonlinear free vibration.

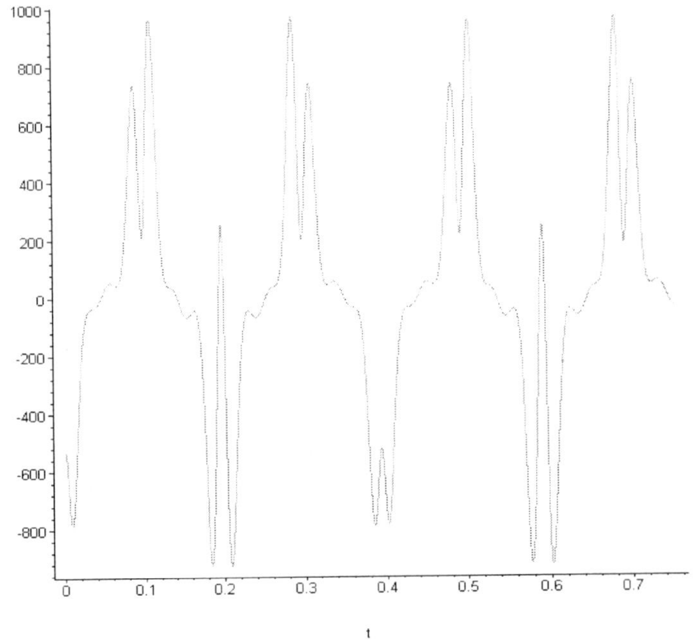

Figure 3.23. Acceleration function, two DOF nonlinear free vibration.

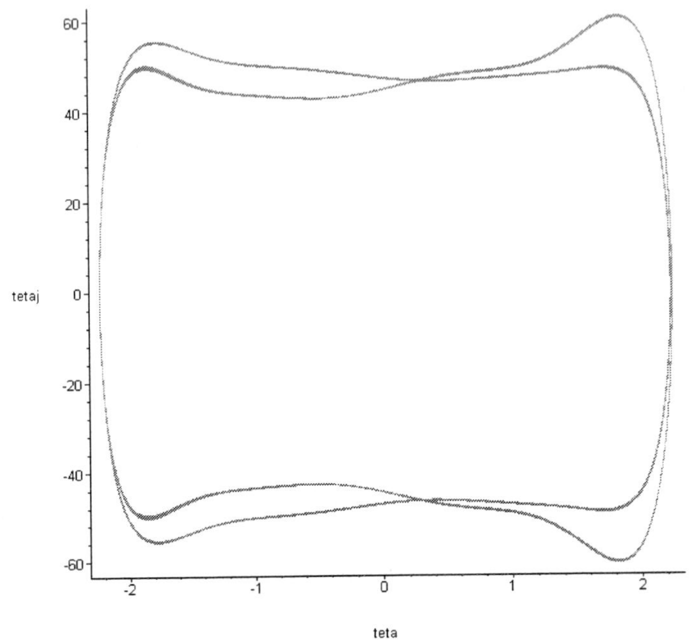

Figure 3.24. Phase diagram, two DOF nonlinear free vibration, angular variable.

Following the same procedure, we first solved the linear free vibration. Figure 3.19 shows the phase diagram, and in this case we found a similar diagram as in Figure 3.1. The main difference comes from the numerical solution when solving both degrees of freedom.

From Figure 3.24, it can be notice that the phase diagram presents shape that is stable.

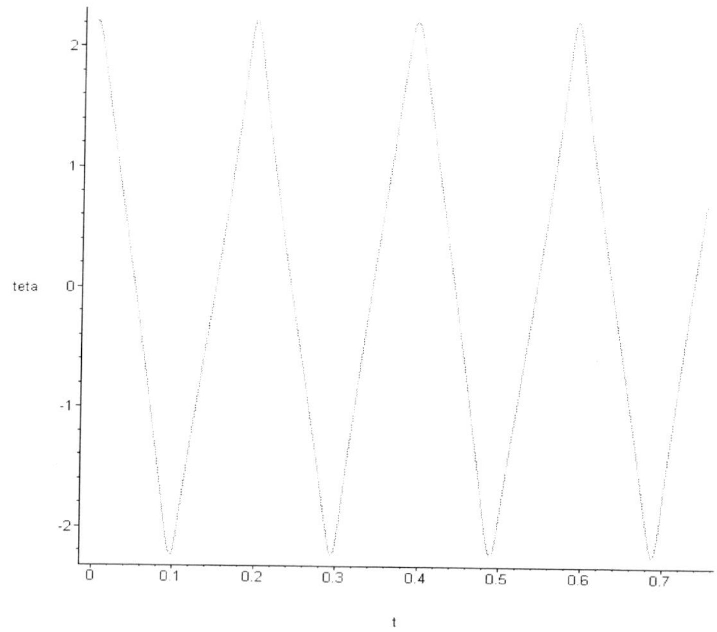

Figure 3.25. Angular displacement, two DOF nonlinear free vibration, angular variable.

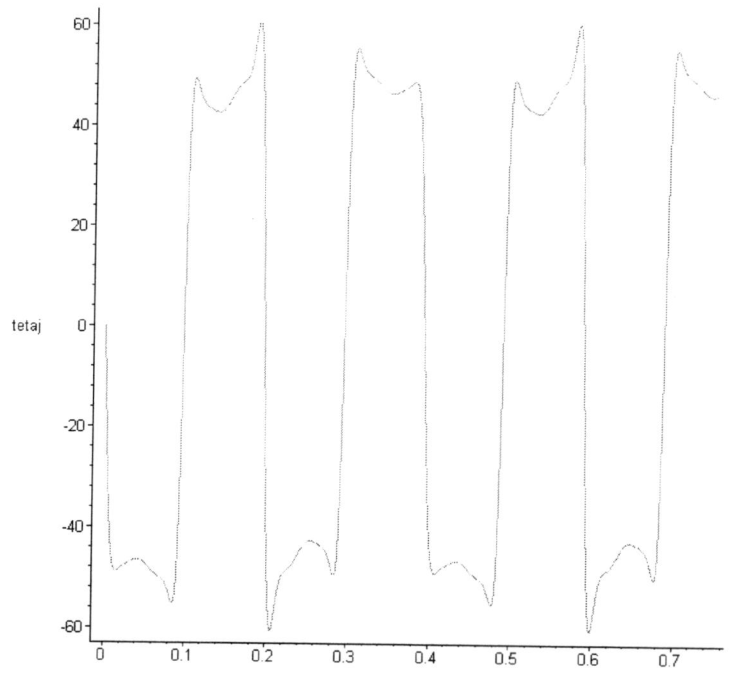

Figure 3.26. Angular velocity, two DOF nonlinear free vibration, angular variable.

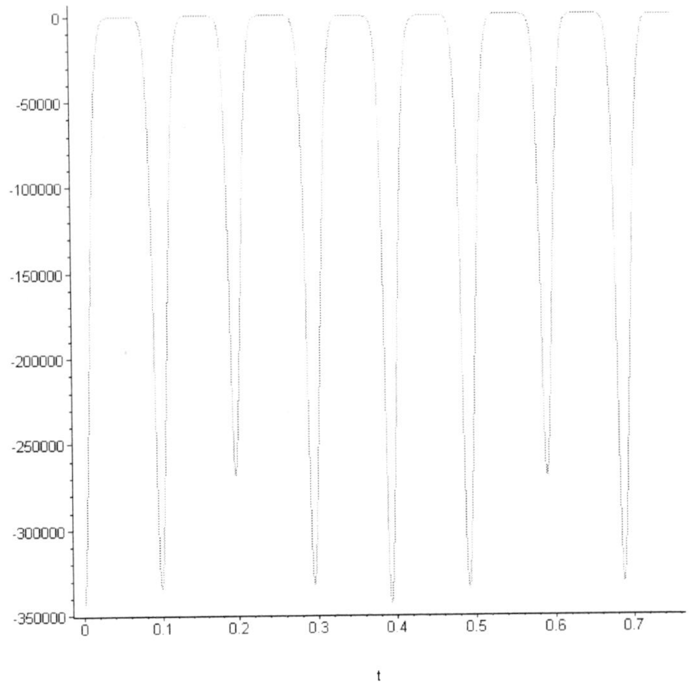

Figure 3.27. Angular acceleration, two DOF nonlinear free vibration, angular variable.

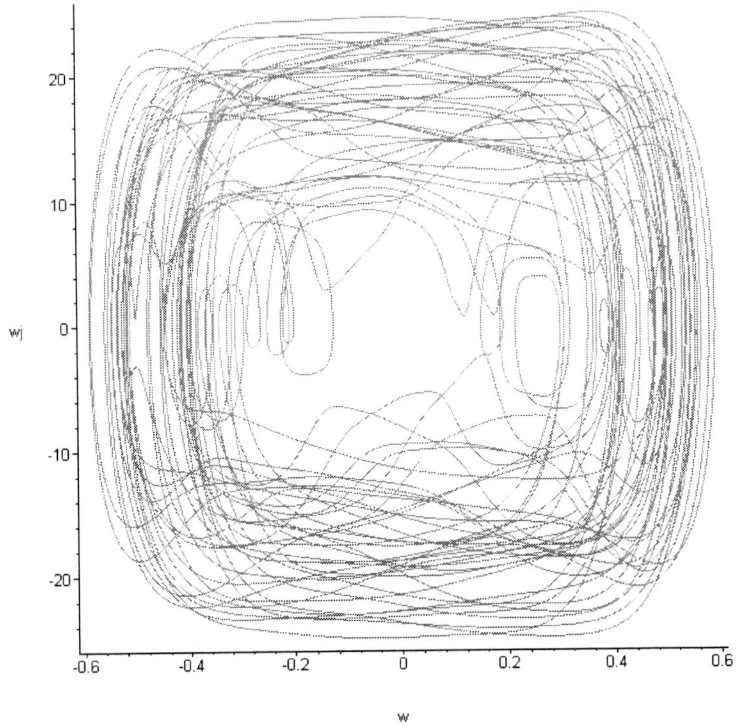

Figure 3.28. Phase diagram. two DOF nonlinear force vibration, at 10% of the nonlinear frequency.

It would be expected that the angular displacement would have the same shape as the linear velocity, but there is a small difference. On the other hand, it is a 90° out phase between them. One of the reasons for the difference between the two solutions is due to the limitations imposed by the numerical solution. It is more difficult to solve the two DOF model, and the stability of the solution method reduces the solution space. But, on the other hand, both analyses give similar results regarding the nonlinear frequency, and the distortion shape of the sinusoidal input function. Thus, it is confirmed that the source of nonlinearity of a slender element is its ability to develop large deformations without having permanent distortions. Both analyses show that the acceleration function is closed to an impulse function.

Similarly as the one degree of freedom, we study the force vibration at different frequencies. The first case corresponds to an exciting frequency equals 10% of the nonlinear frequency.

This system also shows a non sinusoidal wave, with a frequency near 66.84 *rad/s*. We applied an exciting force with this frequency, and the phase diagram shows a similar rectangular pattern as the one degree of freedom system.

Comparing Figure 3.29 with Figure 3.15 it is clear that both systems show a similar behavior under the same exciting force.

The phase diagram shows some discontinuities that reflect the apple shape (Figure 3.24) with a superimposed linear frequency. As in the one degree of freedom case, the system shows no resonance due to the damping effect of the nonlinear terms.

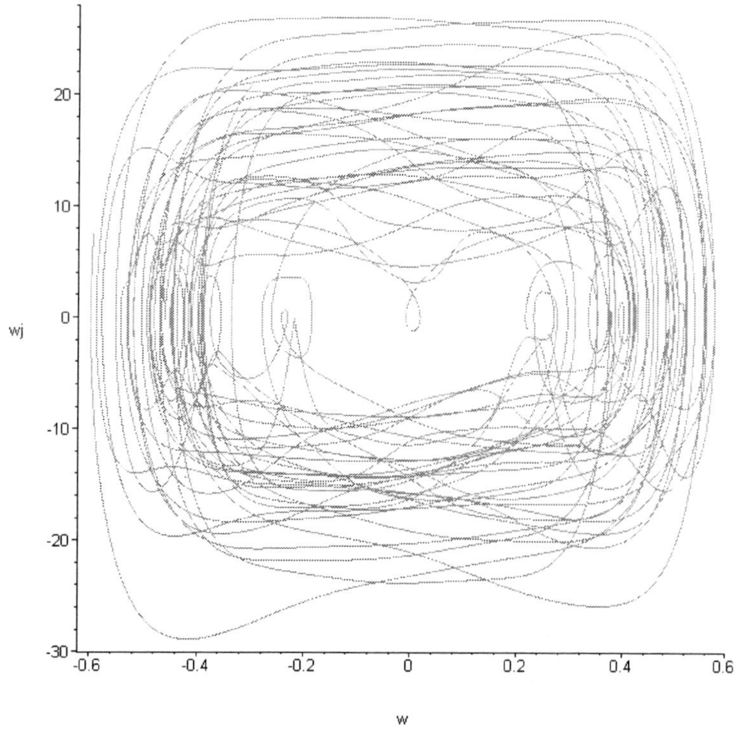

Figure 3.29. Phase diagram. two DOF nonlinear force vibration, at nonlinear frequency.

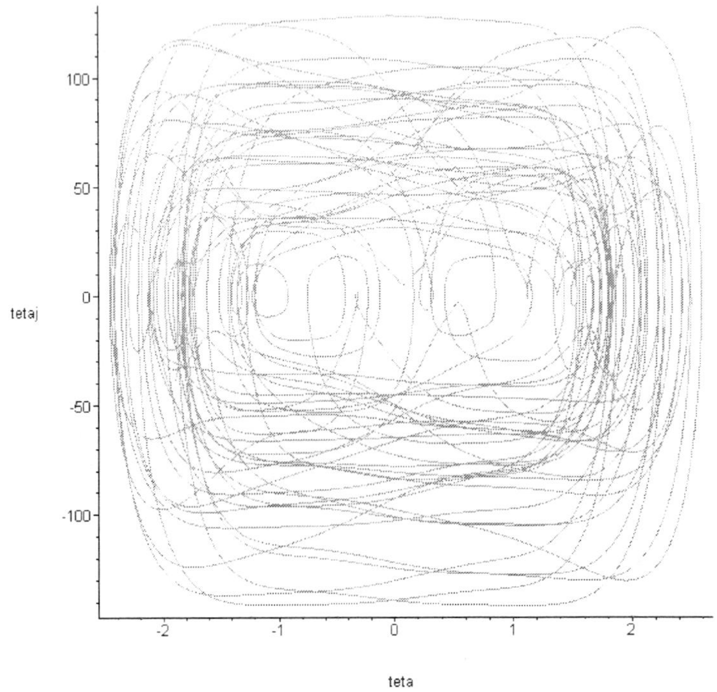

Figure 3.30. Phase diagram. two DOF nonlinear force vibration, angular variable.

Experimental Results

Both models describe the essential behavior of slender beams. It is clear that a slender beam shows a nonlinear response when it is largely deformed. The nonlinearity can be analyzed with a Duffing´s type equation, where the nonlinear terms are a consequence of large bending deformation. It is important to verify the numerical solution; therefore, a very flexible beam was tested and the test results were compared to the numerical results.

The experiment set up consisted of a very flexible steel beam, mounted in cantilever, with an accelerometer glued to the free end. The accelerometer had a minimum weight, thus its contribution to the beam's dynamic response was neglected. The beam was tested under two conditions: In the first test, the beam was largely deformed and it was let free to vibrate. Due to its structural damping, the large deformation vanished at a certain time, but the nonlinear effects were clearly identified at first cycles. In the second test, the beam was forced to vibrate at its maximum amplitude; under this condition, it was also possible to identify the nonlinear effects.

The acceleration data were integrated through an analog device, thus the velocity and displacement were capture with an acquisition system. The wave form of the displacement is shown in Figure 3.31, it can be notice that the wave is not sinusoidal and it has a drastic change of slope near the crest and the valley. This shape is a clear indication of the nonlinear response.

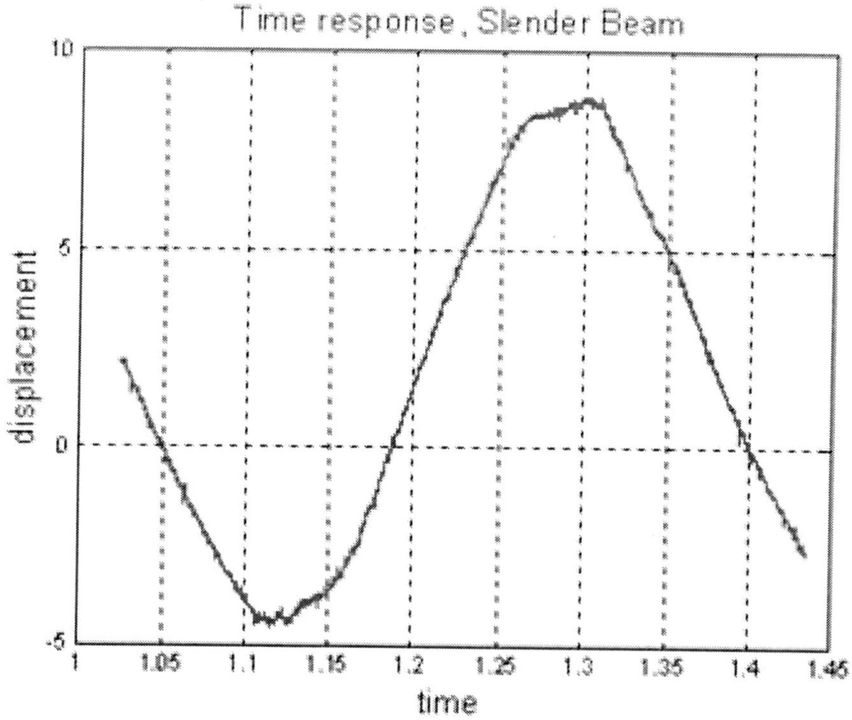

Figure 3.31. Wave form of the displacement of a slender beam under a large deformation initial condition.

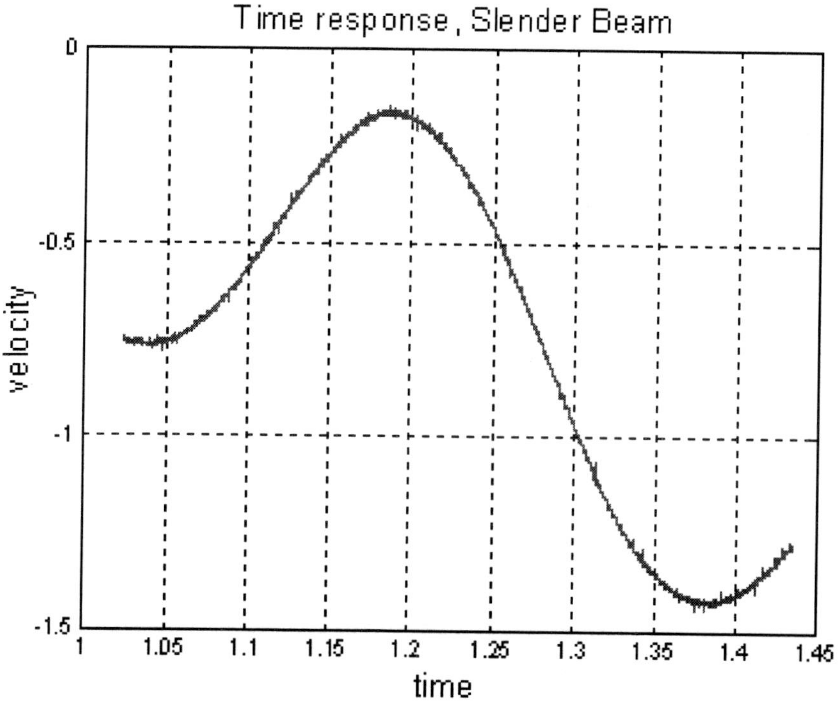

Figure 3.32. velocity wave form of a slender beam.

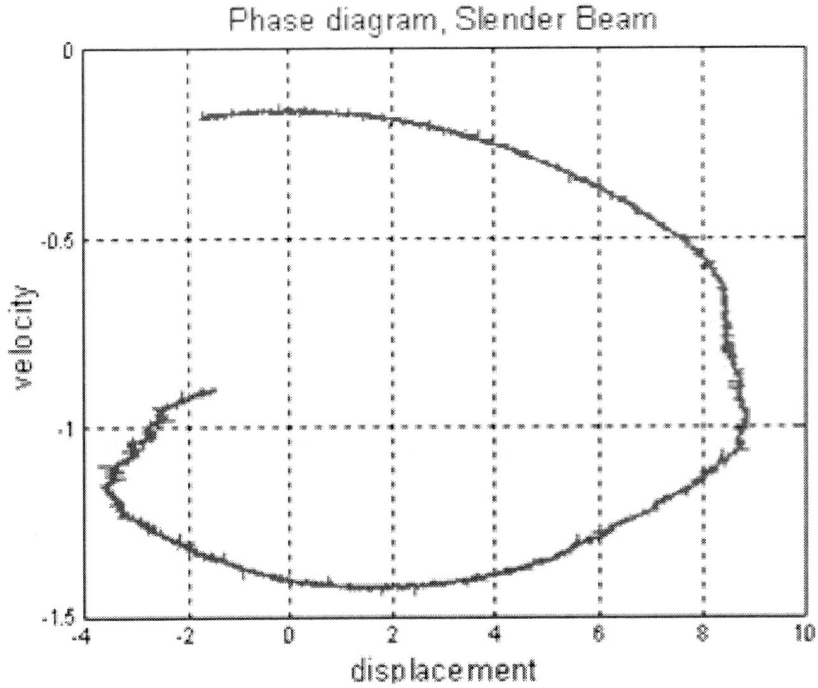

Figure 3.33. Phase diagram of a slender beam due to free vibration.

The wave form of the velocity is plotted in Figure 3.32. In this case the periodic function is periodic but not sinusoidal.

Once the velocity and the displacement function were captured, we plotted the phase diagram, as shown in Figure 3.33. For the free vibration case, the phase diagram reduces its evolving area every cycle, but for the initial loops, it can be identified that the response is rather polygonal than elliptic.

Two major conclusions are drawn: Phase diagram is able to identify the nonlinear response, and the nonlinear effect is caused by the large deformation on the beam.

The second test consisted on a forced excitation of the beam at a frequency where it developed the maximum amplitude. The acceleration was recorded in the same way as in the previous case, and it was also integrated with an analog device. We produce the displacement and velocity plots as shown in Figures 3.34 and 3.35.

It is clear that under this condition, the slope also changes at the crest and valley of the displacement function.

The phase diagram is plotted in Figure 3.36. In this case, the vibration amplitude reflects the excitation force, thus the envelop shape remains almost constant at every cycle, but the nonlinear effect appears as a distortion of the elliptical shape, producing a polygonal figure.

We can conclude that slender beams have a similar behavior as a mechanical system with stiffness discontinuities. They present the same Duffing's type of equation, but, in this case, the coefficients are the result of the large deformation of the beam.

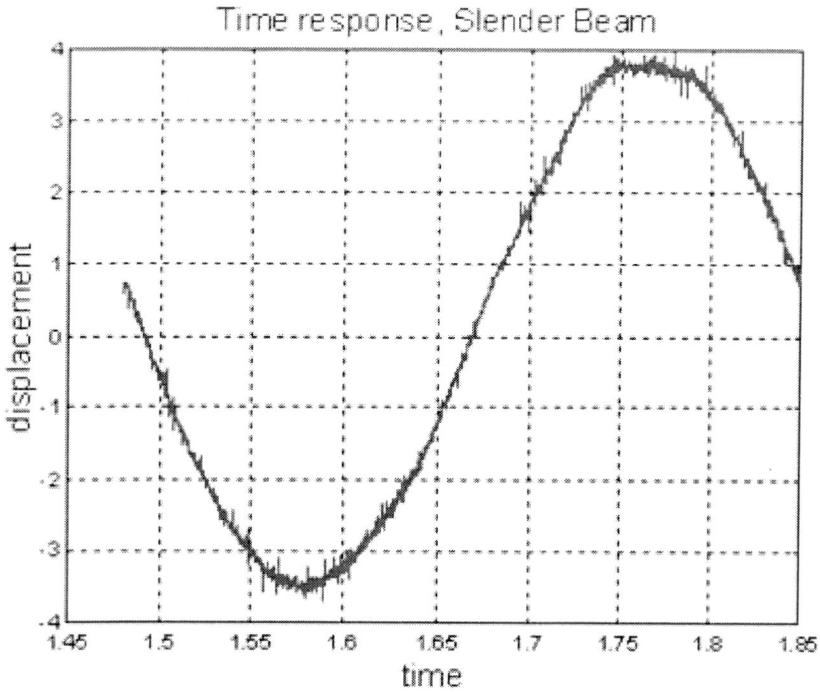

Figure 3.34. Displacement functions of a slender beam under forced vibration.

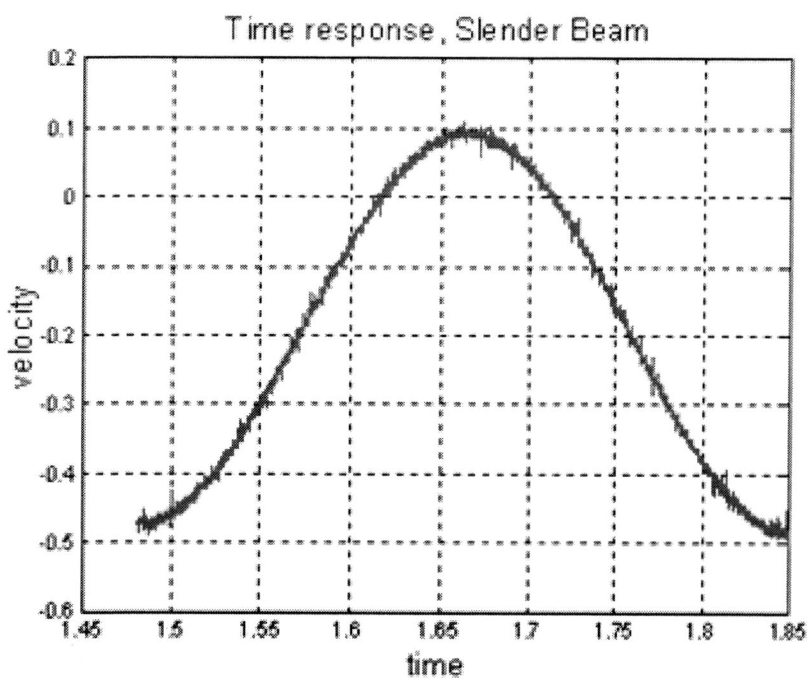

Figure 3.35. Velocity function of the slender beam under forced vibration.

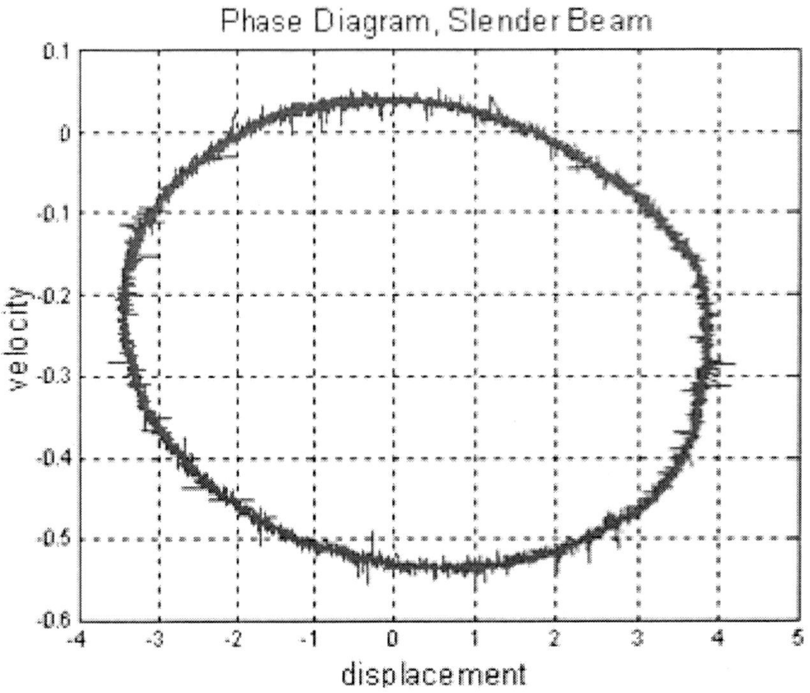

Figure 3.36. Phase diagram of a slender beam under forced vibration.

Chapter 4

MECHANICAL SYSTEMS THAT COMBINE SOFT AND HARD MATERIALS

INTRODUCTION

In Chapter 3 we described the effects of large deformations on the dynamic behavior of solid elements (trust elements) and beams. It was shown that solid elements develop nonlinear responses only when they deform beyond their elastic limit. In many cases this is unpractical or unreal, except for elastomers or similar materials; but for such type of materials, viscoelastic models provide simpler solutions. Soft and hard material combinations are the primary design for passive vibration isolators. Their main function is to separate an object from the vibration source. The need for these devices is increasing because new machinery tends to operate at higher speeds, and manufacturing accuracy is also increasing; even more, noise and vibration tolerances are more stringent. Their proper design is critical in vehicle applications since they are a key factor on passenger comfort.

Passive vibration isolators are a well known solution for mechanical systems that show high vibration levels. Particularly, they help reducing vibrations from engines, presses or other machines with reciprocant mechanism. They combine an elastomeric material with steel metal; the elastomeric material acts as a damper due to the material's internal friction. In their design, the soft material is place between the metal, forming layers in a sandwich type construction.

Another application of layer construction is torsional vibration isolators; they are design using the additional mass concept. With this concept, a free movement ring is mounted to a rotating shaft through a spring. The spring is made of an elastomeric material and it has two dynamic functions: as spring, and as damper. Meanwhile, elastomeric bushings are used in suspension systems of automobiles and other vehicles to reduce shocks and vibrations transmitted from the wheels to the frame of the vehicle. A bushing is a hollow cylinder of elastomeric material contained between an outer cylinder and an inner steel cylindrical rod. In order to predict the dynamic behavior of the suspension system it is important to have a correct model of the bushing. This is, however, complicated due to the elastomeric material in the bushing which is viscoelastic and nonlinear.

Vibration isolation is the most widely used technique for vibration protection. It breaks the transmitted wave by modifying the structural stiffness of the system, and by adding a higher damper.

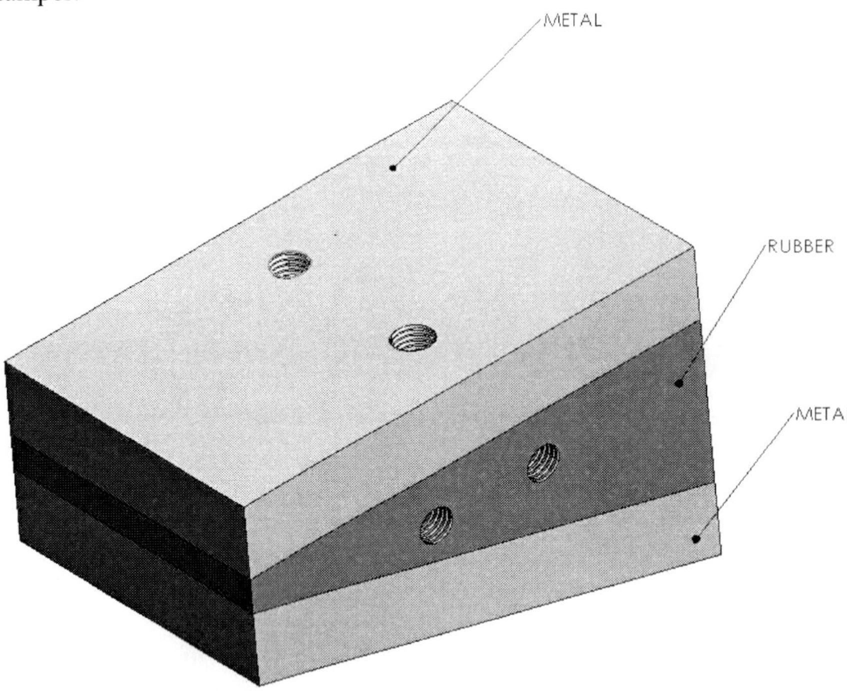

Figure 4.1. Passive isolator.

Since the isolation limitations are more restrictive, more complex designs are required. Dimensions and geometries are far from a simple sandwich earlier design, where an elastomeric plate is placed in between the vibration source and the object. Thus, the basic principle of these devices is to interrupt the vibration wave through a change in stiffness; the easiest way to achieve it, it's through the use of a combination of layers of soft and hard materials.

Vibration isolators are described in a very basic way in text books, and with relatively more detail in handbooks. But more detail information is required, especially when its design requires more complex geometries; even more, with the aid of modern computer tools, the design possibilities increases tremendously.

In this chapter we will present the basis of more sophisticated tools that allows the design of complex mechanical systems, such as those used in the isolation of hazardous vibrations. A complex soft and hard material isolator is placed in between a supporting surface and a vibration source. In most cases, the system is modeled as a set of lump-masses connected through springs and dampers, and the number of degrees of freedom is reduced as minimum as possible. The best way of designing a passive isolator is with the use of the Finite Element Method.

In the Finite Element Method, the continuum is divided into a finite number of regions connected though point or nodes; the regions, or elements, have simple geometry that can easily be integrated. The displacement of any point within the region is interpolated with a polynomial function. The polynomial degree is determined by the number of degrees of freedom, they could be the nodal displacements or nodal slopes. For this purpose, there are many commercial packages for the analysis of kinematics and dynamics of multi-body systems, but few of them deal with vibration analysis in a 3D space; except for some Finite Element Analysis (FEA) packages that include rigid bodies as point masses or point inertias, and add single artificial stiffness elements. This approach is acceptable when the model could represent the actual problem, but it can create ill-conditioned stiffness matrices; thus either it has no solution or its solution is unreal. To avoid this, a special FEA formulation is required and different techniques should be incorporated in the analysis, as well as and adequate matrix construction.

There are other solutions based on the multidimensional model from a Newton-Euler approach, where the partial differential equations have been solved with different finite difference schemes. Most of the work done regarding torsional vibration absorbers solved the problem using finite differences, and they assume a Kelvin type viscoelastic material for the constitutive equation.

In this chapter, we present a Finite Element formulation for multidimensional problems. The multidimensional model starts from a Kelvin type material, the stress-strain relation considers that the linear elastic module affects the volumetric terms and the viscous module affects only the shear terms. This assumption arises from the fact that the viscoelastic model combines the elastic properties of a solid and the viscous behavior of a fluid. Thus, assuming that the viscous module follows a Newtonian behavior, it is only affected by the shear terms.

THE ISOLATOR DESIGN

The basic concept of the isolator is simplified as shown in Figure 4.1. The dynamic parameters are determined from the transmissibility force concept; it can be demonstrated that the amount of force transmitted by a "floating" mass has a magnitude of (nomenclature corresponds to a well know *m-C-K* lumped mass system)

$$|T| = \sqrt{\left(\frac{1+4\zeta^2 r}{(1-r^2)^2 + 4\zeta^2 r^2}\right)} \tag{4.1}$$

Where:

$$\zeta = \frac{C}{2\sqrt{Km}}$$

$$r = \frac{\omega}{\omega_n} \tag{4.2}$$

$$\omega_n = \sqrt{\frac{K}{m}}$$

And ω is the excitation frequency. From these equations, it can be shown that the magnitude of the transmitted force $|T| < 1$ if $r > \sqrt{2}$. That means that the excitation frequency most be grater that the natural frequency. But in many cases this type of design is not possible and the operating frequencies are a lot lower that the natural frequencies. Thus, the stiffness and damping characteristics should be modified by adding a soft material between the vibration source and the object that will be isolated. Since there are two design parameters, the problem is under determined; therefore, a range of frequencies and an amplitude threshold must be defined. With these values, a set of damping and stiffness values can be determined from Figure 4.2. (Each curve corresponds to a particular damping factor ζ)

Once the stiffness and damping coefficients are determined, the design process iterates within several geometries and different material properties. Must of the isolators are built from elastomeric materials, thus in order to include this properties in the FEA formulation, an adequate damping matrix must be established. We present a damping matrix formulation based on viscoelastic model.

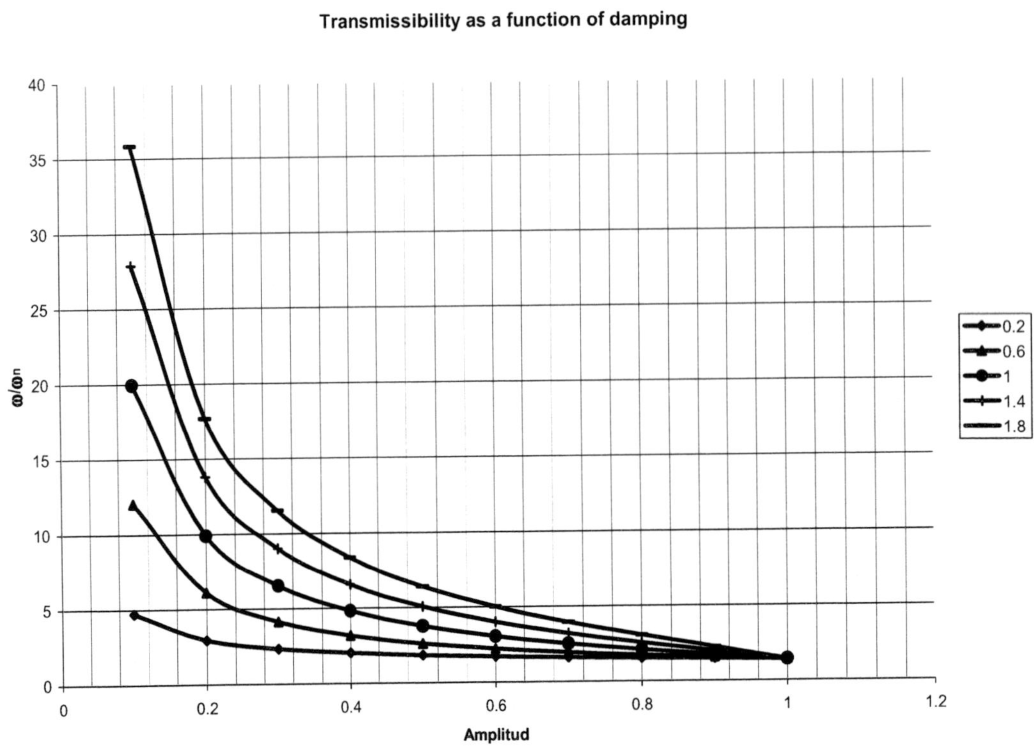

Figure 4.2. Transmissibility factor of a passive isolator.

VISCOELASTICITY

In previous chapters, we presented mechanical systems that present a nonlinear stiffness; in this chapter we define a design method for vibration isolators. In Chapter 3, we introduce the effects of large deformations on solid elements, and we showed its limitations for practical applications. As we said before, only elastomeric type material will present large deformations before yielding or breaking, but we also explained that simpler formulations provide better solutions for practical purposes. This is the reason why we present this method here. The method is based on a Finite Element approach for combine materials, where a rigid part (metal) is linked through a soft material (elastomer). The damping properties are derived assuming that the elastomer behaves as a Kelvin type viscoelastic material, and it can be described as a simple system with a linera spring in parallel with a linear viscous damper. A linear spring will show a relationship with its deformation as:

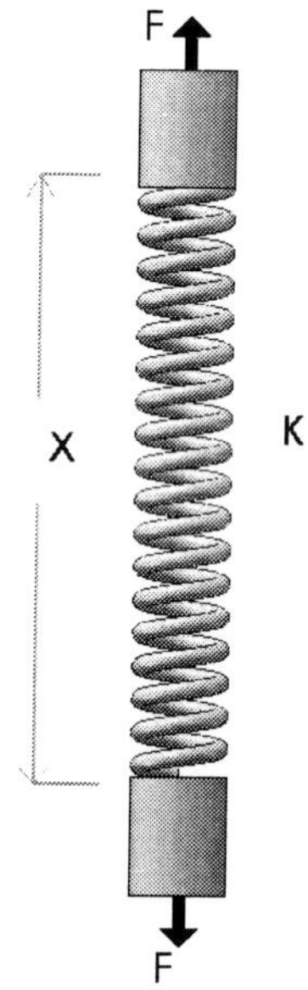

Figure 4.3. Spring model.

$$F = kx$$
$$\sigma = E\varepsilon \tag{4.3}$$

On the other hand, viscous fluids have part the stress proportional to the rate of strain:

$$F = \eta\dot{x}$$
$$\tau = \mu\dot{\varepsilon} \tag{4.4}$$

Such as shock absorbers and dampers. Polymers and metals at high temperatures exhibit more complicated mechanical behavior; they simultaneously show elastic deformation and viscous dissipation. By combining spring and dampers either in series or in parallel different material properties can be simulated.

For the most common rubber materials employed in isolators, the well known Kevin-Voigt model can be applied to represent its behavior.

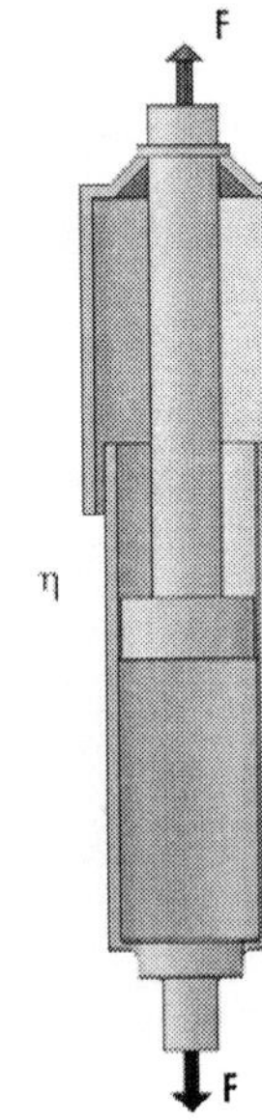

Figure 4.4. Viscous damper.

Thus, the constitutive equation for a one-dimensional material is formed combining the damping effect with the spring

$$\sigma = E\varepsilon + \mu\dot{\varepsilon} \tag{4.5}$$

This model does not allow for instantaneous response but is simple enough to model passive vibration isolators. A more general form of the constitutive equation can be expressed as:

$$\underset{\sim}{\tau} = 2e\underset{\sim}{\varepsilon} + \lambda(tr\,\varepsilon)\underset{\sim}{I} + 2\mu\underset{\sim}{\dot{\varepsilon}}\lambda_d(tr\,\dot{\varepsilon})\underset{\sim}{I} \tag{4.6}$$

Separating the shear effect from the volumetric deformation and using the FEA notation, the constitutive equation can be expressed as:

$$\{\sigma\} = [E]\{\varepsilon\} + [\mu']\{\dot{\gamma}\} \tag{4.7}$$

The equations of motion are derived from the Hamilton's principle. Although Hamilton's principle and Lagrange's equation are equivalent, Hamilton's principle is broader in the sense that it applies to distributed systems. In this case, it is assumed that it holds for an individual element as:

$$\frac{d}{dt}\left(\frac{\partial L^e}{\partial u_{ti}}\right) - \frac{\partial L^e}{\partial u_i} = 0 \tag{4.8}$$

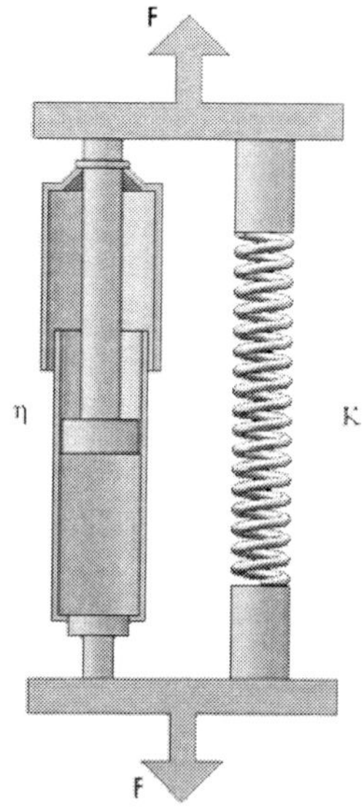

Figure 4.5. Viscoelastic model.

Where u_i is the displacement function evaluated at a nodal point, and u_{ti} is the velocity function at the same nodal point, and:

$$L^e = T^e - W^e \tag{4.9}$$

The kinetic energy T^e for each finite element is calculated as:

$$T^e = \frac{1}{2}\int_V \rho\{\dot{u}\}^T\{\dot{u}\}dV \tag{4.10}$$

The other terms in Hamilton's principle combine two terms: The required work to deform the element (internal energy), and the work done by the external forces.

$$W^e = W_E^e - W_I^e \tag{4.11}$$

The deformation energy can be calculated as:

$$W_I^e = \frac{1}{2}\int_V \{\varepsilon\}^T\{\sigma\}dV \tag{4.12}$$

Substituting the constitutive equation (Eq. 4.7) into the deformation energy term:

$$W_I^e = \frac{1}{2}\int_V \{\varepsilon\}^T[E]\{\varepsilon\}dV + \frac{1}{2}\int_V \{\varepsilon\}^T[\mu']\{\dot{\gamma}\}dV \tag{4.13}$$

Applying a finite element approach, the field displacement can be approximated by a linear polynomial,

$$u(x) = \sum_{i=1}^{n} a_i x^{i-1} \tag{4.14}$$

Or, in a matrix form:

$$u(x) = [N]\{u\} \tag{4.15}$$

Where, x represents the space coordinates vector and $\{u\}$ represents nodal displacements. The norm of $\{u\}$ is equal to the number of degrees of freedom of each element.

The strain and the strain rate functions can be expressed in terms of the polynomial coefficients as:

$$\begin{aligned} \{\varepsilon\} &= [B]\{u\} \\ \{\dot{\gamma}\} &= [A]\{\dot{u}\} \end{aligned} \tag{4.16}$$

In this way, the deformation energy is expressed as:

$$W_I^e = \frac{1}{2}\int_V \{u\}^T [B]^T [E][B]\{u\}dV + \frac{1}{2}\int_V \{u\}^T [B]^T [\mu'][A]\{\dot{u}\}dV \tag{4.17}$$

The work done by the external forces depends upon a distributed pressure f on the element surface, and the generalized nodal loads:

$$W_E^e = \int fudx + \sum_{1}^{m} Q_i^e u_i dx \tag{4.18}$$

Thus, Hamilton's principle is written as:

$$\frac{d}{dt}\left(\frac{\partial T^e}{\partial u_{ti}}\right) + \frac{\partial W^e}{\partial u_i} = 0 \tag{4.19}$$

Therefore, the equation of motion of a single element can be simplified as:

$$[M]^e\{\ddot{u}\} + [C]^e\{\dot{u}\} + [K]^e\{u\} = \{f\} \tag{4.20}$$

Where:

$$
\begin{aligned}
[M]^e &= \int_V \rho [N]^T [N] dV \\
[C]^e &= \frac{1}{2} \int_V [B]^T [\mu'][A] dV \\
[K]^e &= \int_V [B]^T [E][B] dV \\
\{f\} &= \int_l f dx + \{q\}
\end{aligned}
\tag{4.21}
$$

The solution of the equation of motion can be obtained by applying a finite difference scheme, but in the case of the mechanical systems that we are studying here, the mass and stiffness matrices are ill-conditioned. Therefore, it is necessary to modify the matrices in order to find a solution for practical cases. For this purposes, the Guyan reduction is applied. The model is reduced by removing those degrees of freedom that are associated with a relatively small mass. If we defined a partition vector $\{u_2\}$ for those nodes to be removed; then, the transformation equation can be rearrange as:

$$
\begin{bmatrix} M_{11} & M_{12} \\ M_{21} & M_{22} \end{bmatrix} \begin{Bmatrix} \ddot{u}_1 \\ \ddot{u}_2 \end{Bmatrix} + \begin{bmatrix} C_{11} & C_{12} \\ C_{21} & C_{22} \end{bmatrix} \begin{Bmatrix} \dot{u}_1 \\ \dot{u}_2 \end{Bmatrix} + \begin{bmatrix} K_{11} & K_{12} \\ K_{21} & K_{22} \end{bmatrix} \begin{Bmatrix} u_1 \\ u_2 \end{Bmatrix} = \begin{Bmatrix} f \\ 0 \end{Bmatrix}
\tag{4.22}
$$

If we assume that sub-matrices $M_{22} =0$, and $M_{12} =0$, and harmonic motion, the partition vector can be defined as

$$
\begin{aligned}
\{\ddot{u}_2\} &= -[C_{22}]^{-1}[C_{21}]\{\dot{u}_1\} \\
\{u_2\} &= -[K_{22}]^{-1}[K_{21}]\{u_1\}
\end{aligned}
\tag{4.23}
$$

Thus, the new equation of motion can be expressed as:

$$
[M_{11}]\{\ddot{u}_1\} + \left([C_{11}] - [C_{12}][C_{22}]^{-1}[C_{21}]\right)\{\dot{u}_1\} + \left([K_{11}] - [K_{12}][K_{22}]^{-1}[K_{21}]\right)\{u_1\} = \{f\}
\tag{4.24}
$$

With this arrangement, the ill-condition matrices are recomposed, and the solution will avoid numerical errors.

Unidirectional Model

Although there are many references for elastomer properties, there are few methods for determining their viscoelastic coefficients as a Kelvin type material. The unidirectional model can be applied for the experimental determination of viscoelastic properties. If we applied the same procedure to a trust element with one degree of freedom per node, the equation of motion is derived as follows.

The polynomial coefficients in terms of nodal displacement are:

$$a_1 = \frac{1}{l}(u_1 x_2 - u_2 x_1)$$
$$a_2 = \frac{1}{l}(u_2 - u_1) \tag{4.25}$$

where u_1 represents the displacement of the first node, and u_2 represents the displacement of the other node.

Thus, the form matrix is:

$$u(x) = \frac{1}{l}\left[(x_2 - x) \quad (x - x_1)\right]\begin{Bmatrix} u_1 \\ u_2 \end{Bmatrix} \tag{4.26}$$

The strain and strain rate vectors will be:

$$\{\varepsilon\} = \frac{1}{l}\left[-1 \quad 1\right]\begin{Bmatrix} u_1 \\ u_2 \end{Bmatrix}$$
$$\{\dot{\gamma}\} = \frac{1}{l}\left[-1 \quad 1\right]\begin{Bmatrix} \dot{u}_1 \\ \dot{u}_2 \end{Bmatrix} \tag{4.27}$$

Substituting these matrices into equations (4.21), the equation of motion becomes:

$$[M]^e = \begin{bmatrix} m & 0 \\ 0 & 0 \end{bmatrix}$$
$$[C]^e = \frac{\mu' A}{2l}\begin{bmatrix} 1 & -1 \\ -1 & 1 \end{bmatrix} \tag{4.28}$$
$$[K]^e = \frac{EA}{l}\begin{bmatrix} 1 & -1 \\ -1 & 1 \end{bmatrix}$$

If we consider only a single degree of freedom, the equation of motion is simplified as:

$$m\ddot{u} + \frac{\mu' A}{2l}\dot{u} + \frac{EA}{l}u = f(t) \tag{4.29}$$

It is equivalent to the well known vibration equation

$$m\ddot{u} + c\dot{u} + ku = f(t) \tag{4.30}$$

where

$$c = \frac{\mu' A}{2l}, k = \frac{EA}{l}$$

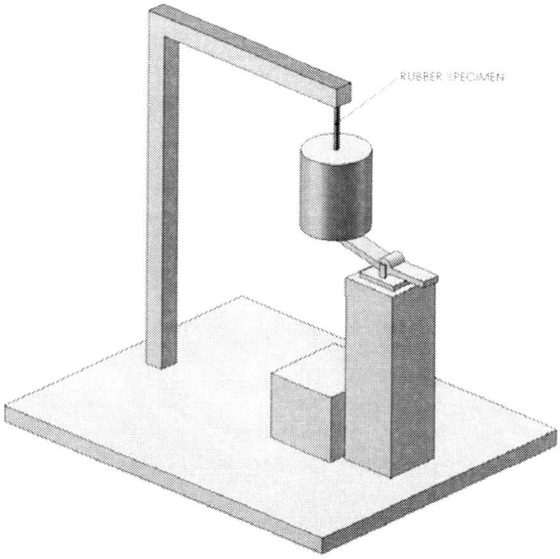

Figure 4.6. Viscoelastic test rig.

According to ASTM D-5992-96, we can determined the viscoelastic properties of a elastomeric material, by measuring the free vibration response of a mass attached to elastomeric strip of length l and cross section area A (Fig. 4.6). The vibrations are measured with a test rig that has a rigid frame where the elastomeric strip suspends the mass. The mass is displaced and then it is let to freely vibrate. With a proximity probe, the actual displacements of the mass are recorded with a data acquisition system. Then; the elastic and viscous properties of the elastomeric strip are automatically calculated with the logarithm decrement principle:

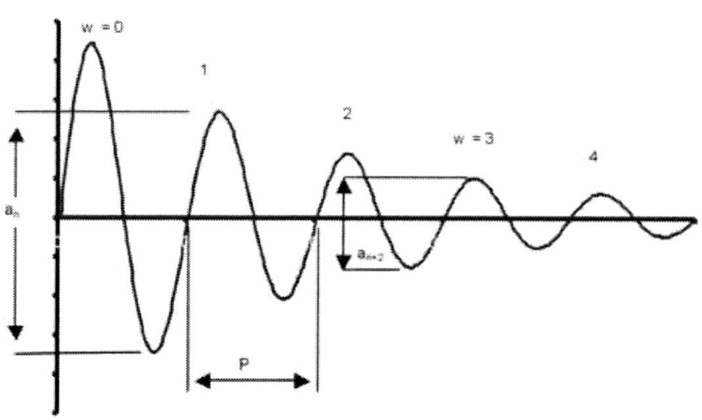

Figure 4.7. Displacement measurements of a rubber specimen.

The stiffness of the strip is found using the frequency measurement:

$$k = \frac{4\pi^2 m}{P^2} = 2\pi^2 f^2 m \tag{4.31}$$

And, from equation (4.29), the elastic modulus is calculated as:

$$E = \frac{kl}{A} \tag{4.32}$$

From the logarithm decrement, the damping ration is estimated as:

$$\zeta = \frac{1}{\sqrt{1 + \frac{4\pi^2}{\delta^2}}} \tag{4.33}$$

where

$$\delta = \frac{1}{i}\ln\left(\frac{a_n}{a_{n+i}}\right) \tag{4.34}$$

The damping constant c is related to the damping ratio as:

$$c = 2\zeta\sqrt{km} \tag{4.35}$$

Comparing equations (4.29) and (4.30), the viscous term can be calculated as:

$$\mu' = 4\zeta\sqrt{\frac{Elm}{A}} \tag{4.36}$$

Since the viscoelastic parameters are calculated from experimental data, a statistical management needs to be included. The actual value is estimated as:

$$x = \overline{x} \pm 1.96\frac{s}{\sqrt{N}} \tag{4.37}$$

Where x is the actual value, s is the standard deviation and N is the number of measurements. Applying this procedure to a rubber strip, the typical viscoelastic values for this material are:

E	5.0 ± 0.032 MPa
μ'	65.0 ± 0.13 KPa-s

With the adequate viscoelastic parameters, the design of passive isolators depends upon the dynamic parameters. Damping and stiffness will depend fundamentally on the geometry of the elastomeric material, and the mass will depend on the metallic elements that will compose it.

Bidirectional Model

Following the same procedure, we present a bidirectional model. This model can be applied to plane stress, plane strain or axis-symmetric problems. The difference among them are $[E]$ and $[\mu']$ matrices that are defined for each type of analysis. For a linear element, the interpolating polynomial has the form:

$$\begin{aligned} u(x,y) &= a_0 + a_1 x + a_2 y \\ v(x,y) &= b_0 + b_1 x + b_2 y \end{aligned} \tag{4.38}$$

Expressing the coefficients with the matrix notation

$$\begin{Bmatrix} a_0 \\ a_1 \\ a_2 \end{Bmatrix} = \frac{1}{\Delta} \begin{bmatrix} N_1 & N_2 & N_3 \\ N_4 & N_5 & N_6 \\ N_7 & N_8 & N_9 \end{bmatrix} \begin{Bmatrix} u_i \\ u_j \\ u_k \end{Bmatrix} \tag{4.39}$$

and

$$\begin{Bmatrix} b_0 \\ b_1 \\ b_2 \end{Bmatrix} = \frac{1}{\Delta} \begin{bmatrix} N_1 & N_2 & N_3 \\ N_4 & N_5 & N_6 \\ N_7 & N_8 & N_9 \end{bmatrix} \begin{Bmatrix} v_i \\ v_j \\ v_k \end{Bmatrix} \tag{4.40}$$

Where

$$\Delta = (x_j y_k - x_k y_j) - (x_i y_k - x_k y_i) + (x_i y_j - x_j y_i) \tag{4.41}$$

And

$$
\begin{aligned}
N_1 &= x_j y_k - x_k y_j \\
N_2 &= x_k y_i - x_i y_k \\
N_3 &= x_i y_j - x_j y_i \\
N_4 &= y_i - y_k \\
N_5 &= y_k - y_i \\
N_6 &= y_i - y_j \\
N_7 &= x_k - x_j \\
N_8 &= x_i - x_k \\
N_9 &= x_j - x_i
\end{aligned}
\tag{4.42}
$$

Finally, the form matrix will have the form

$$
[B] = \frac{1}{\Delta}\begin{bmatrix} N_4 & 0 & N_5 & 0 & N_6 & 0 \\ 0 & N_7 & 0 & N_8 & 0 & N_9 \\ N_7 & N_4 & N_8 & N_5 & N_9 & N_6 \end{bmatrix}
\tag{4.43}
$$

Similarly, the $[A]$ matrix can be obtained. The stiffness and damping matrices are obtained following the same steps as in the unidirectional example. After developing the element matrices, the integration and matrix reduction depends on each particular problem, and in many commercial programs it is done automatically (Eq. 4.21). Next we will present an example of a torsional isolator.

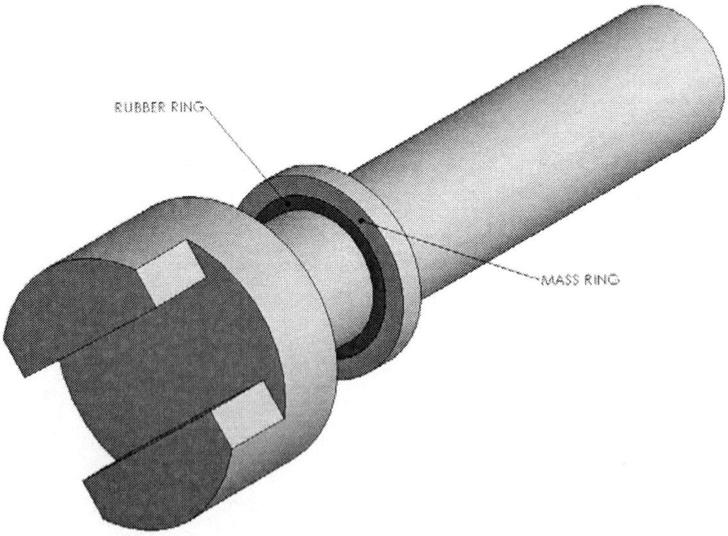

Figure 4.8. Example of a torsional damper application.

TORSONIAL VIBRATION DAMPER

A torsonial damper utilizes the same transmissibility principle describe before. The basic idea consists on adding a mass, which in this case will be a ring circumscribing the shaft, connected through a spring and a damping device. The outer inertia ring is fixed relative to the mounting hub through a ring of elastomeric material. As we discussed before, the elastomeric material by itself has the damping and spring capabilities. These units must be balanced after assembly to correct imbalance due to any offsets between the inertia ring and the hub that are created by the imposition of the rubber ring between them.

The torsonial isolator must be design in such a way that it is tuned to vibrate out-of-phase with the fundamental torsional natural frequency of the rotating system. The properties of the elastomer allow it to absorb the vibration, converting the mechanical energy into heat energy to achieve damping. The limited capacity of the torsonial isolator could be that it is overheated; otherwise, it could result in a very short life. In addition, the main rotating assembly to which the isolator is tuned includes the crankshaft, pistons, rods, flywheel, and any rotating mass that may be directly coupled or geared to the shaft. Care must be taken in order to avoid that the torsonial isolator resonates with any of the natural frequencies of the rotating elements; this could be catastrophic instead of reducing torsonial vibrations. Since the inertia ring is directly connected to the rotating system through the elastomer, the power supply will have an additional load and will create higher stress as it accelerates. The additional loading would be directly proportional to the inertia of the ring.

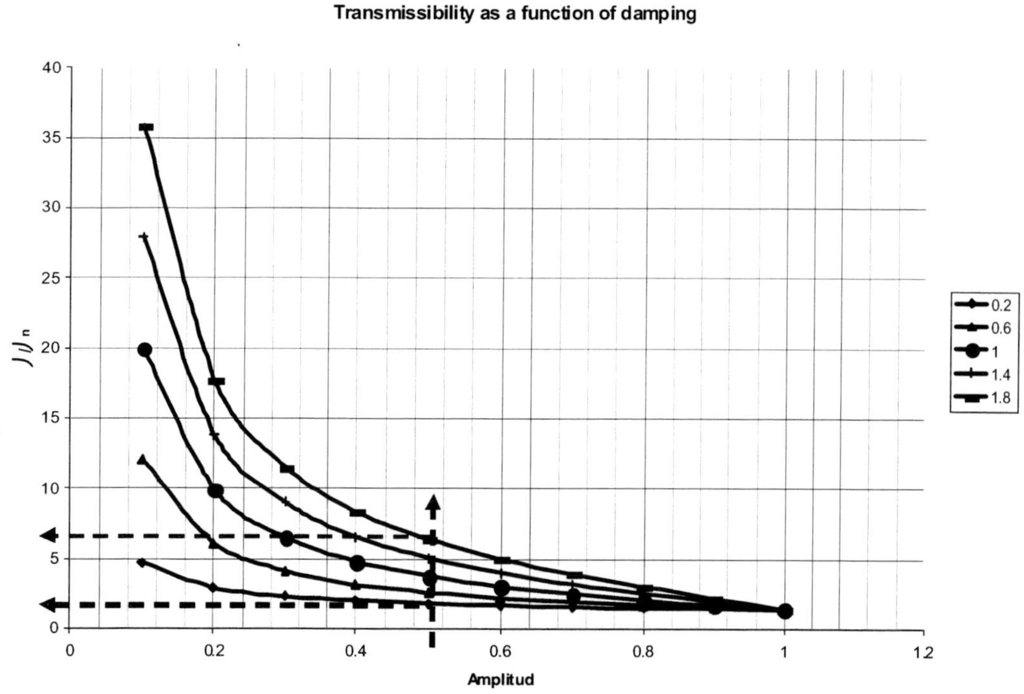

Figure 4.9. Torsional damper example.

The design procedure starts with the definition of the range of frequencies that need to be reduced. The isolator will work mainly at this range; it is important to review all the principal natural frequencies of all the elements that integrate the rotating system. Then, it the designer must specify a threshold value for the transmissibility torque; in other words, the designer must set the T value. With this parameter, the range of natural frequencies and damping factors can be selected as shown in Figure 4.9.

From this figure, the damping factor must be between 0.2 and 1.8, and the frequency range between 1.5 and 5.3.

In this case, the frequency that needs to be reduced is at 250 Hz. Thus, the natural frequency of the sandwich metal-elastomeric ring should be:

Natural Frequency	Hz
ω_n	47.1
	167

With these values, there are several solutions. To begin with, it is necessary to select the mass for the metal ring; in this case 0.6kg ring was chosen. Then, from the associated damping factors and the required natural frequencies, the rubber ring's stiffness and damping are:

Solution	K (KN/m)	C (Ns/m)
1	58.1	76.56
2	730.7	2506.9

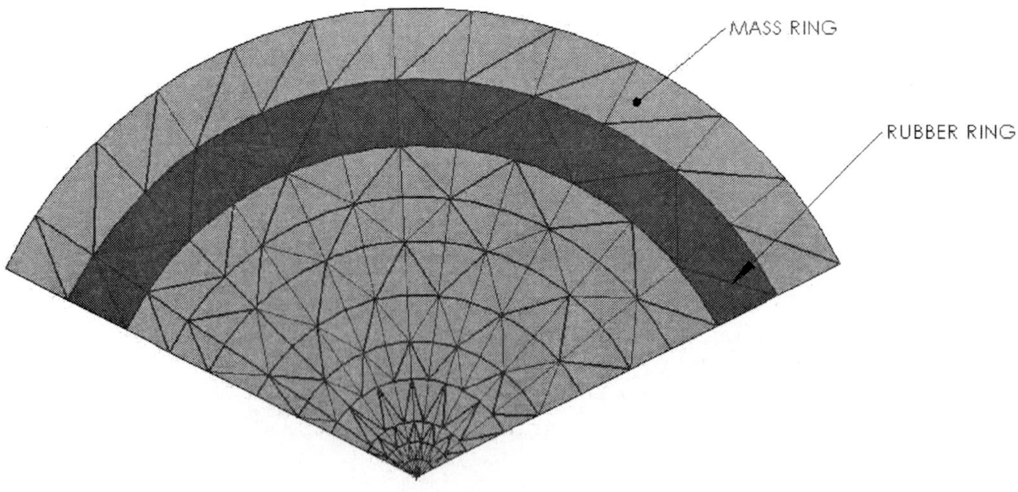

Figure 4.10. Finite element model of a torsional isolator section.

These values must be obtained with the finite element model. If we select a rubber such as the test specimen presented before; then, there are two undetermined values, the thickness of the rubber ring and its width.

After iterating among different geometries, the final design of the torsional isolator is presented in Figure 4.10.

In this way, the design of systems that combine soft and hard materials could be accomplished by applying appropriate matrix reduction techniques, and the incorporation of modified damping matrices.

Chapter 5

SOLUTION METHODS

INTRODUCTION

In previous chapters, we have presented different mechanical systems that are modeled through nonlinear differential equations. These models were also discussed in chapter 1, and we have presented different approaches to solve them. For a linear system, the solution of the problem is very straight forward, once the mathematical model is been defined, there are many algorithms to find the answer. Whereas, in nonlinear systems; the solution method plays an important role on the solution of the problem. Therefore, the selection and definition of the solution algorithm is critical.

In this chapter we describe the most useful techniques that identify nonlinear behaviors on mechanical systems with discontinuities. For these systems, the spectral and modal analyses lack the ability to identify transient and nonlinear responses; thus, the most adequate techniques are:

- System identification through harmonic distortion
- System stability through phase diagram
- Transient response through time-frequency maps

Modern machinery is design with tougher conditions: high speed combined with lighter structures. With this combination, new machines become more competitive than older design. Intelligent manufacturing systems required on line monitoring and control systems. Traditionally, monitoring systems are based on the Fast Fourier Transform (FFT), which, due to its basis is unable to identify transient responses, and nonlinear behaviors. On the other hand, the FFT requires a considerable processing time that limits its application to early fault detections. The development of new sensors, signal processing techniques and faster microprocessors are key elements for modern monitoring systems. These systems require a better understanding of the machine response and the nature of the output signals. In this chapter we present some of the basic solution methods for identifying the behavior of transient and nonlinear systems. In this chapter we describe: The concept of harmonic distortion, the construction of the phase diagram for common signals, and the construction of time-frequency maps.

In order to improve monitoring systems, other analysis techniques are needed. In consequence, monitoring systems should include the techniques we are describing next.

SYSTEM IDENTIFICATION

Harmonic Distortion

In Chapter 3, we have shown that slender beams have a nonlinear structure, but its nonlinear behavior appears only at certain conditions. These conditions depend upon their geometry, the excitation frequency and the initial conditions. Thus, in practice, we must have the ability to establish if the beam is vibrating in a nonlinear pattern or not. One way of identifying it is through a modal test, following a linear superposition procedure. If this principle is violated, errors will appear in the data. The definition of a linear system is the superposition principle, once a system behaves differently from the principle it can be classified as a nonlinear system. From previous discussions, it is clear that a slender beam with large deformations violates the superposition principle. It is clear how the phase diagram changes from an ellipse (Chapter 3) to a rectangular shape. If the system would hold the superposition principle, the phase diagram would be elliptical.

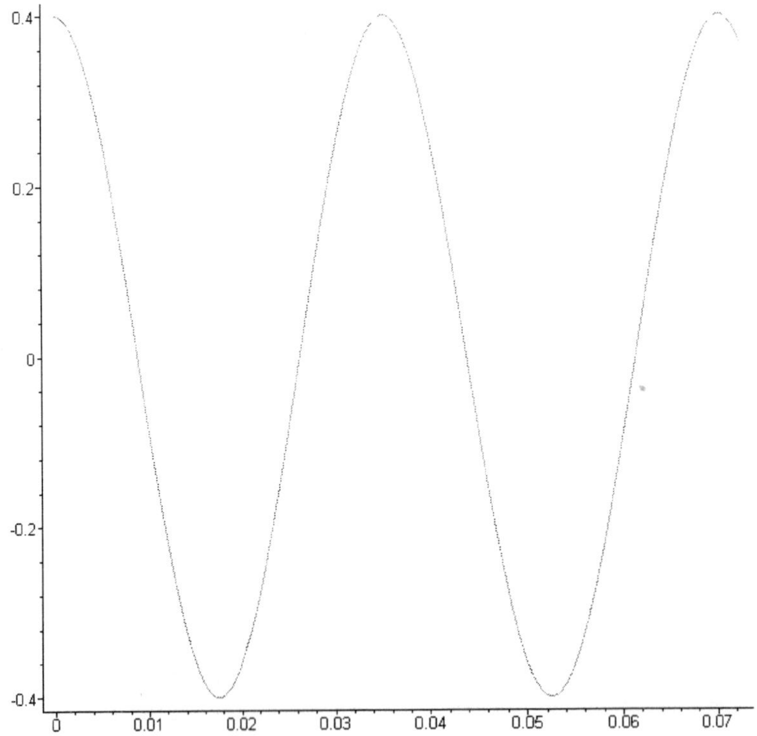

Figure 5.1. Sinusoidal input signal.

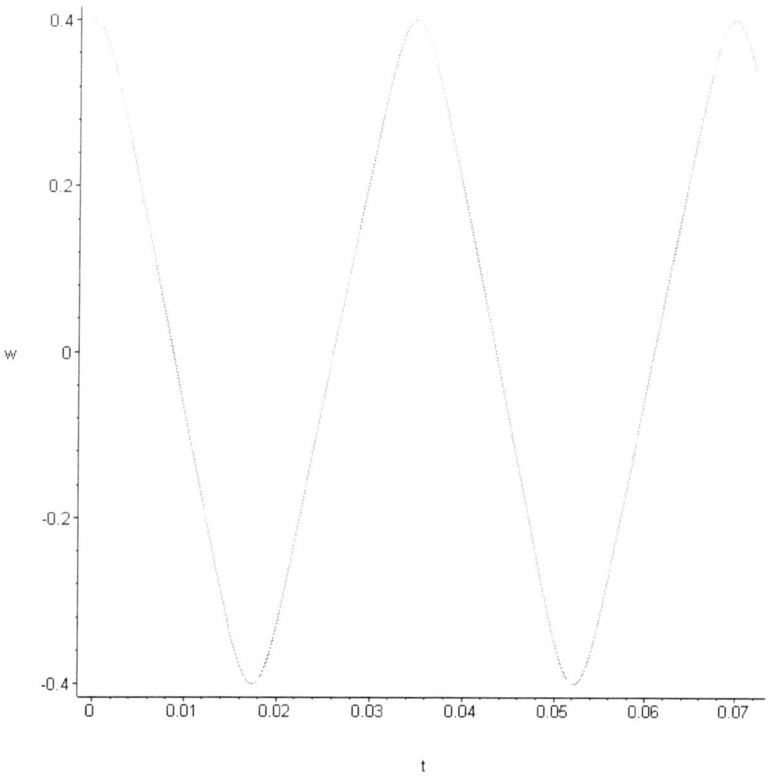

Figure 5.2. Triangular output signal.

Harmonic distortion is one of the clearest indicators of nonlinearity. It is a straightforward consequence of the principle of superposition. If the excitation force is sinusoidal the beam displacement must be sinusoidal with the same frequency. Distortion is easily identified at any of previous figures, where the displacement functions are harmonic but not sinusoidal. Thus, one way of identifying a mechanical system with nonlinear behavior is through the application of sinusoidal excitation and comparing the "shape" of the output displacement, if it is not sinusoidal, then the system is nonlinear. In practice is not that simple. Other signal analysis techniques, such as wavelets, can identify this "shape" variation.

Homogeneity

Another technique to test if a system behaves nonlinearly is through reviewing the homogeneity and distortion of the transfer function in the frequency domain.

Homogeneity means that the transfer function will be invariant under changes of the level of excitation. This is not always true, since there are some nonlinear systems, which show homogeneity. The reason is that this property is a weaker condition than superposition. Only if a control system is employed to maintain a constant excitation force spectrum, nonlinearity can be detected through homogeneity.

Reciprocity

The other important property, which can be used to detect nonlinearities, is reciprocity. When employing reciprocity all parameters must be the same, e.g. the output must be acceleration and the input forces. Thus, if it is a linear system, the stiffness matrix will be symmetric, otherwise it will be a nonlinear system.

PHASE DIAGRAM ANALYSIS

A practical method for identifying nonlinear responses is the phase diagram. In practice, the phase diagram represents the relationship between the output signal and its first time derivative, in a parametric form. In this way, the time variable is eliminated from the diagram. Phase diagram analysis can be implemented electronically with an analog integration circuit.

If we have an output signal such as:

$$x = X\cos(\omega t + \phi) \tag{5.1}$$

its first time derivative will be:

$$\dot{x} = -X\omega sen(\omega t + \phi) \tag{5.2}$$

Then, its phase diagram will be expressed as:

$$S = S(x, \dot{x}) \qquad \forall t$$

If we plot *S*, we will get an ellipse as shown in Figure 5.3 (for clarity, the scales are modified)

One of the great advantages of the phase diagram is that it can show if a system is linear or not. As it has been discussed in previous chapters, nonlinear systems create a non-smooth phase diagram, with unstable patterns and unpredictable shapes. Knowing this, a phase diagram can be utilized for predicting discontinuities in the stiffness of mechanical systems. An ideal linear system will produce a phase diagram with a single ellipse at any time, and will be independent of its initial conditions. Thus, the slope of the function $S = S(x, \dot{x})$ (Chapter 1) will be a tangent function as shown in Figure 5.4. Whereas, the out put shape of a nonlinear system will vary drastically. Therefore, any sign change within a period loop, will be a clear indication of an instable condition. This is a fundamental condition for the application of the phase diagram as a monitoring tool.

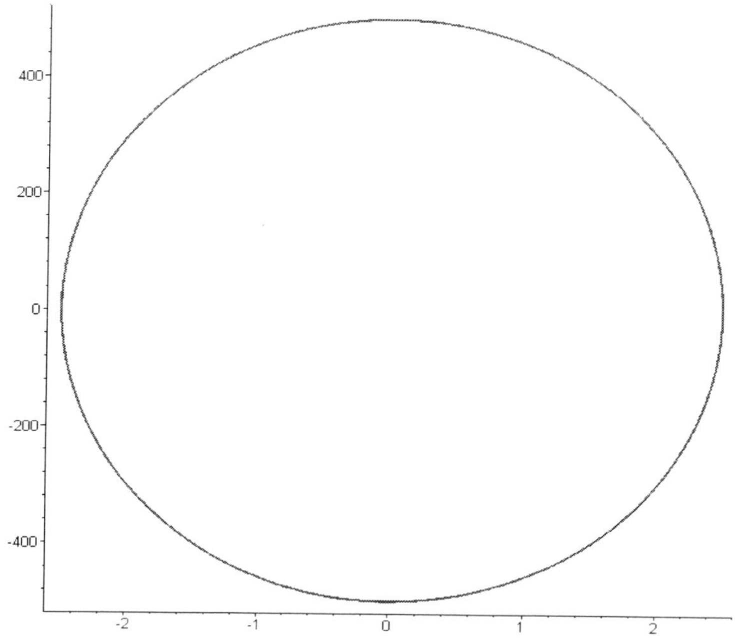

Figure 5.3. Phase diagram of a sinusoidal output signal.

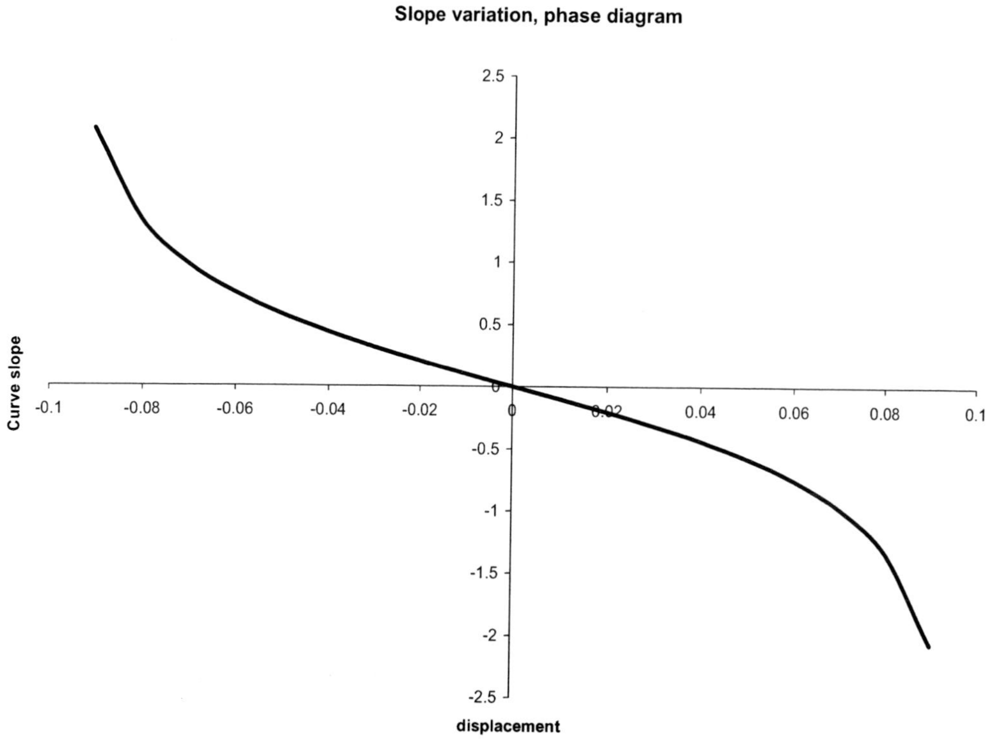

Figure 5.4. Variation of the phase diagram slope, sinusoidal output signal.

The phase diagram is sensible to the combination of output signals. A system with a dominant sinusoidal output signal will produce an elliptical shape. This figure will be stable at any period T of the dominant signal (similar to the one shown in Figure 5.3)

Every output signal has a characteristic phase diagram. We include some of them as reference. Figure 5.5 shows the phase diagram of an output signal with a function such as:

$$x = a\cos(\omega_1 t + \phi_1) + b\cos(\omega_2 t + \phi_2) \tag{5.4}$$

where

$$a = b$$

$$\omega_2 > \omega_1$$

$$\phi_1 = \phi_2$$

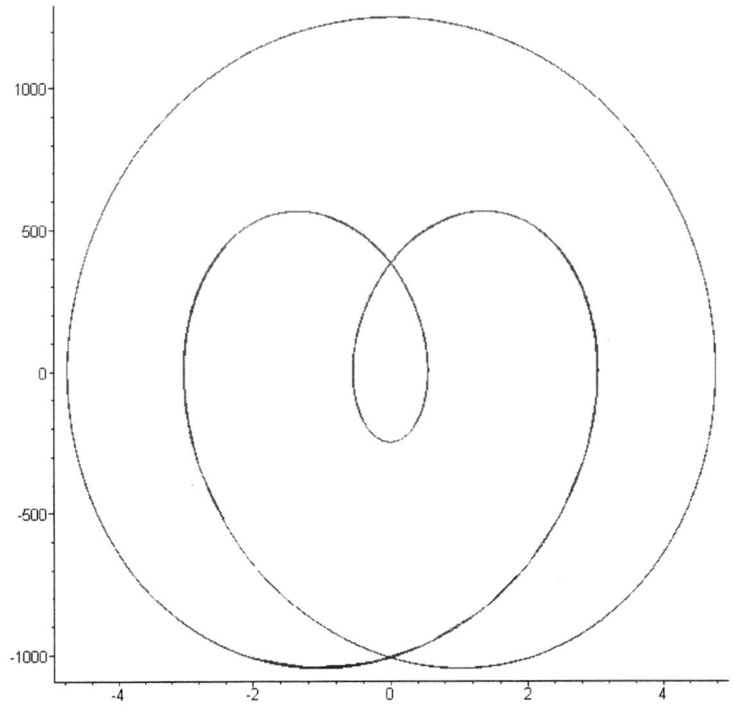

Figure 5.5. Phase diagram of an output signal $a=b$, $\phi_1=\phi_2$.

If the output signal has two dominant frequencies with different amplitudes and the same phase angle, its phase diagram will look like the one shown in Figure 5.6.

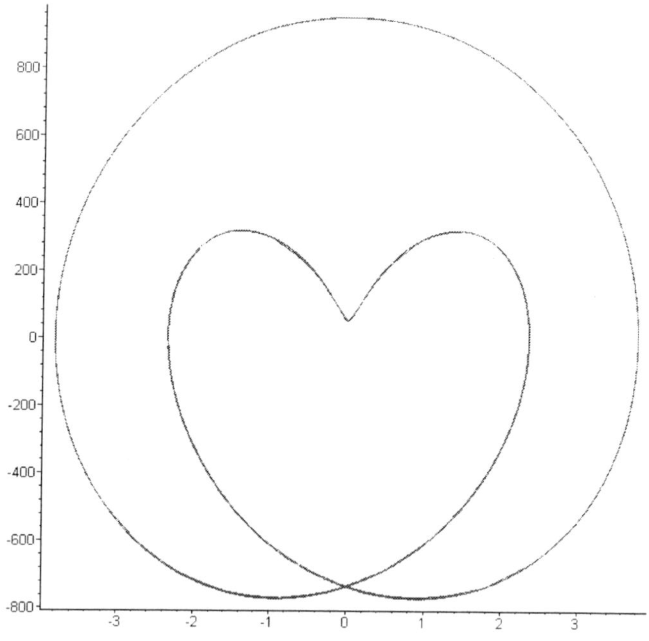

Figure 5.6. Phase diagram of an output signal $a>b,\ \phi_1=\phi_2$.

But if they are out of phase 180°, then the phase diagram also rotates 180° (Figure 5.7)

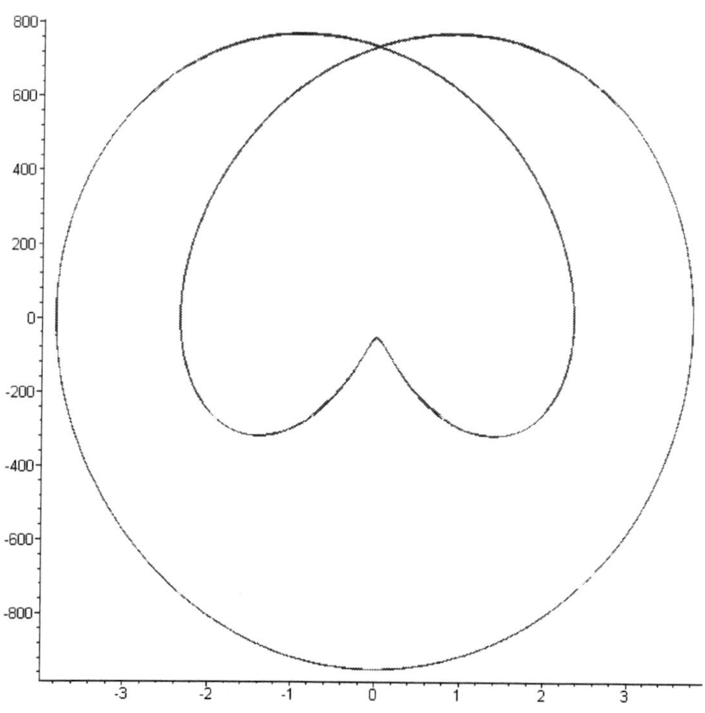

Figure 5.7. Phase diagram of an output signal $a>b,\ \phi_1= 180°+\phi_2$.

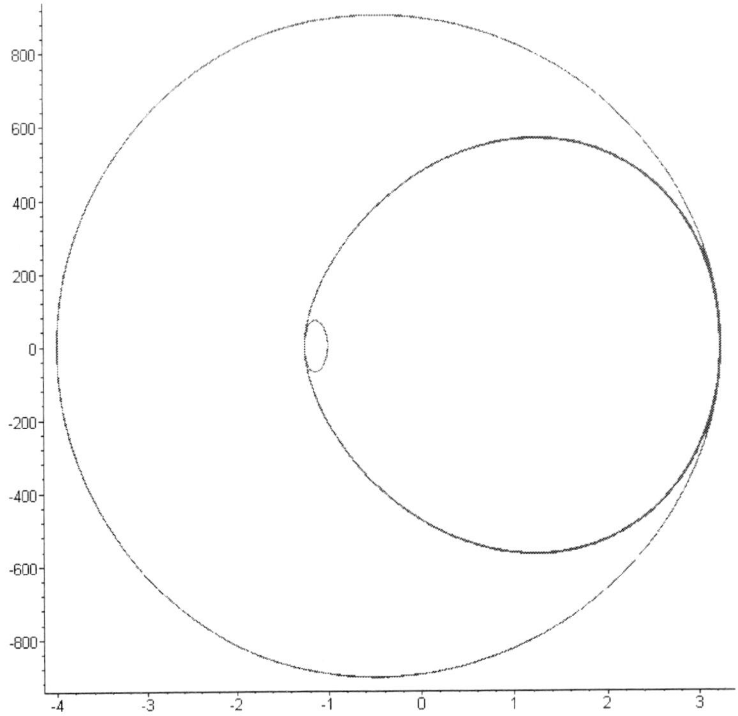

Figure 5.8. Phase diagram of an output signal $a>b$, $\phi_1=90°+\phi_2$.

When the two dominant frequencies have different amplitudes and are out of phase 90°, the phase diagram has the form shown in Figure 5.8.

Another common output signal is a fundamental frequency modulated by a high frequency signal. (For example, the response of a pair of damaged gears). Fig 5.9, shows a phase diagram of an output signal with low modulation, and Figure 5.10 shows the same signal with high modulation. It is important to notice that the number of ellipse varies with the ratio between the high frequency to the low frequency signals.

TIME-FREQUENCY ANALYSIS

Through out the book, we have presented the effects of stiffness discontinuities; in most cases, the output response is nonlinear, and it may have a transient component. In previous chapters, we have applied different signal analysis; these analyses help the understanding of the original system, and they also are useful for the configuration of monitoring systems. In order to simulate a nonlinear output signal we can approximate it by a linear term, a nonlinear term and white noise.

$$x(t) = x_n(t) + x_{nl}(t) + wn(t) \quad (5.6)$$

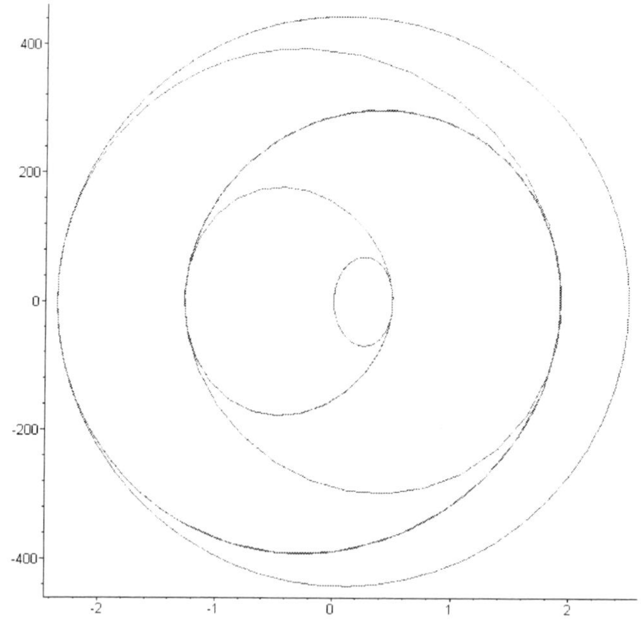

Figure 5.9. Signal with low modulation.

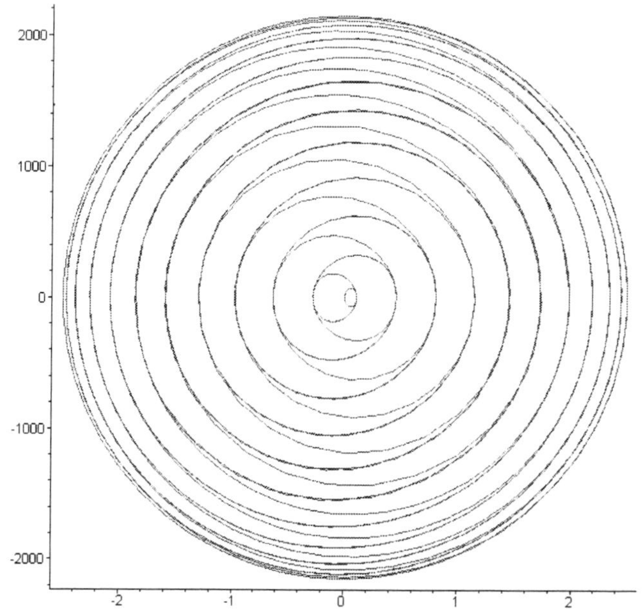

Figure 5.10. Signal with high modulation.

The linear term can be expressed as a Fourier series, white noise as a random non-periodic term and the nonlinear term as:

$$x_{nl}(t) = \sum_{n=1}^{N} a_n \delta_\tau e^{-b_n(t-\tau)} \cos(\omega_n (t-\tau) + \phi_n) \tag{5.7}$$

where

$$\delta_\tau = \begin{cases} 1 & t-\tau \geq 0 \\ 0 & t-\tau < 0 \end{cases} \tag{5.8}$$

This equation includes most of the common transient output signals. Coefficients a and b depend upon the number of transient functions that are present in the output signal. We include it as a reference and, since we can determine the coefficients, the delay parameter τ, and the phase angle ϕ, we can use it as a way of demonstrating the effectiveness of the signal analysis methods.

This output signal has additional information unable to be identified by a simple spectral analysis. The Fourier Transform eliminates the transient part, and it is unable to analyze the nonlinear terms, because it is a linear transformation and holds the superposition principle, whereas the nonlinear signal shows a harmonic distortion. Therefore, the most adequate way of analyzing this type of signals is with a Time Frequency Map.

In a Time Frequency Map, a one dimension signal is decomposed into a two dimension vector. In this way, the time dependent components are separated from the steady state components, and both the transient components and the nonlinear effects can be identified as part of the entire map.

It will be shown in next section the similitude between eq. (5.7) and the STFT.

Short Time Fourier Transform

The first method for producing a Time Frequency Map is The Short Time Fourier Transform (STFT). It is derived from the Wigner probability distribution. Wigner, in 1932, constructed his function over the phase space in quantum mechanics. Later, Ville presented a work for the identification of an instantaneous spectrum. His work was based on Gabor's signal analysis transform. One of the problems of their approach arises from the fact that Wigner's distribution is nonlinear, and it causes many interference phenomena. In order to solve these difficulties, another approach was suggested. This approach cuts the signal into slices, followed by performing a Fourier transform of each slice. In this way, the signal is separate by time intervals (slices) and the frequency components are calculated piecewise. This segmentation produces a non periodic function, which creates discontinuities at the boundaries; thus, the Fourier transform crates large coefficients at high frequencies. To avoid these effects, the windowing concept is introduced. Instead of a rectangular segmentation, a smooth window-function is employed to segment the original signal. The window function is almost one at the origin and zero at the edges. Common window functions are identified with names such as: Hamming, Hanning, Bartlett or Kaiser. If the window function is infinitely differentiable, the localized Fourier coefficients will show a polynomial decay.

We consider unnecessary to include the definition of the Fourier Transform, and the Fast Fourier Transform. We refer to them as a commonly use signal analysis techniques, and we recommend the reader to review their principle in order to understand why they cannot process the transient and nonlinear terms of an output signal. The Short Time Fourier Transform (STFT) is one of the first signal analysis techniques that are able to process the transient part of a signal. What it does, is the transformation of a single dimension vector into a two dimension vector. The method defines a sliding window $w(t)$ centered at time τ. With this function, the Short Time Fourier Transform processes localized frequency analysis. In this way, the analysis identifies time variations on the frequency response.

If we store the signal content, (acceleration or velocity measurements) in a vector $y(t)$, the STFT is computed as:

$$S(\tau, f) = \int w(t-\tau) y(t) e^{-j2\pi ft} dt \tag{5.9}$$

where τ is related to the size of the window function.

The resolution of the signal decomposition depends directly on the window function. Plotting S as a function of t and f will produce a Time-Frequency Map.

The time frequency map is also identified as "spectrogram". With it, we can see how the frequency content of a signal changes over time. A simple explanation of a map construction is: we cut the signal into blocks and compute the spectrum of each block; the blocks can overlap in order to avoid artificial peaks, as explained before. Each block is multiplied by the window function that is tapered at its endpoints.

Several parameters must be chosen:

- Block length
- Type of window function.
- Amount of overlap between blocks
- The block length is determined by the window function $w(t)$

A long block length will provide higher frequency resolution because the window function will be narrow. Meanwhile, a short block length will provide higher time resolution because there are less average calculations at each block. Thus, the higher frequency time-frequency map is identified as *narrow-band spectrogram*, and the higher time time-frequency map is named as *wide-band spectrogram* (short window function). There are other considerations that should be taken into consideration. It is important to put emphasis that the STFT of a signal is invertible.

To do a numerical evaluate the Short Time Fourier Transform, we sample the frequency axis ω in N equally spaced samples from ω=0 to $\omega=2\pi$.

If we transform the continuous function (5.9) into a discrete form, then the STFT function will be

$$S(\tau, f) = \sum_{n=-\infty}^{\infty} y(n) w(n-mR) e^{-j2\pi fn} \tag{5.10}$$

where $\tau = mR$, and R is the block length, $y(n)$ is the discrete signal vector, and w is the discrete window function.

If the window function has a constant overlap property, the sum of all the individual Discrete Time-Function Transform will be the Discrete Time-Function Transform of the entire signal. Thus, the window function can be expressed as:

$$\sum_{m=-\infty}^{\infty} w(n-mR) = 1, \forall n \tag{5.11}$$

and the Discrete Time-Function Transform can be calculated as:

$$\begin{aligned}\sum_{m=-\infty}^{\infty} S(m,f) &= \sum_{m=-\infty}^{\infty} \sum_{n=-\infty}^{\infty} y(n)w(n-mR)e^{-j2\pi f} \\ &= \sum_{n=-\infty}^{\infty} y(n) \sum_{m=-\infty}^{\infty} w(n-mR)e^{-j2\pi f}\end{aligned} \tag{5.12}$$

then

$$\sum_{m=-\infty}^{\infty} S(m,f) = \sum y(n)e^{-j2\pi f} = S(f) \tag{5.13}$$

Then, the window functions that satisfy eq. (5.11) will determine the existence of artificial spectrum peaks. For example, with a Blackman window function (rectangular function), the time results will look all correct, but the data near the edges of the window function will be smoother than the data closer to the center of the window. Another example of a window function will be the Gaussian function:

$$w(t) = e^{-at^2\tau^2} \tag{5.14}$$

where a is constant and τ controls the window width.

However the STFT has also its disadvantages, such as the limit in its time-frequency resolution capability, which is due to the uncertainty principle. Practically, this condition implies that every signal requires a particular window function, and different analysis need to be done before setting the best parameters. If we don't know in advance the type of information we are looking for, the application of this method will require many iteration in order to set the proper window function. Low frequencies can be hardly identified with short windows, whereas short pulses can only poorly be localized in time with long windows.

The STFT has another limitation that is its high redundancy. It can be reduced if we alter the condition the data sampling. Normally, the signal vector is evenly discretized with a finite number of points, such that $t=na$, and the frequency discretization can be done if $f=mb$. Where n and m are integers, and $a,b > 0$. Modifying a and b different Time-Frequency Map grids are produced. These limitations in the resolution are one of the reasons why the wavelet transform is an alternative for vibration analysis.

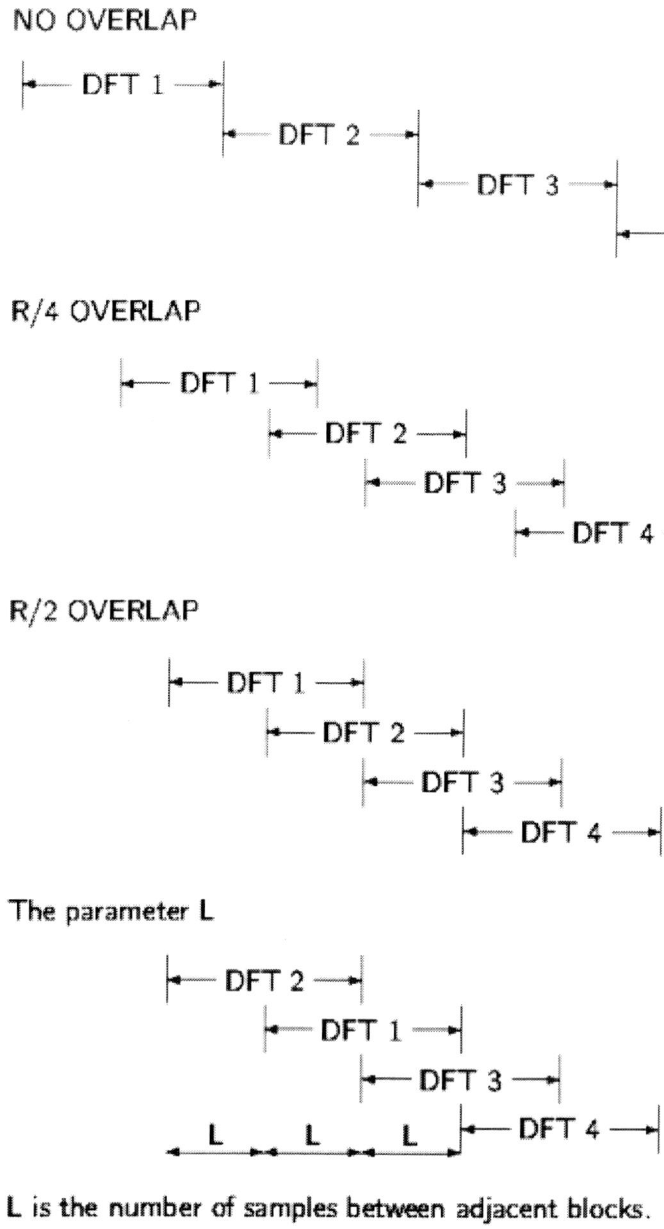

Figure 5.11. Different overlap schemes.

Wavelet Transform

Wavelets have been applied in different fields; they have proven to be a powerful tool for signal analysis and data processing. Although, it is the best transform for many vibration problems, its practical application hasn't been very popular. Most of the commercial analysis systems are based on the FFT, and only specialized literature describe the way wavelets can analyze transient and nonlinear vibration problems.

The big difference between The Wavelet Transform and the STFT is that wavelets can introduce a window function with variable size; thus, instead of a and b been constant values, they can vary depending on the frequency component. Thus, for low frequencies we apply a large window; and, for high frequencies a small window with the same wavelet function. The result is a constant relative-bandwidth analysis which allows us the identification of transient and nonlinear responses that varies with time. The advantage of The Wavelet Transform is that it decomposes the signal into a set of functions at different scales and shifted along the time variable.

In mathematical terms, a wavelet is a function with zero average:

$$\int_{-\infty}^{\infty} \psi(t)dt = 0 \tag{5.15}$$

This is the basic condition for determining the wavelet function, which is dilated by a scale parameter s, and translated by τ:

$$\psi_{\tau,s}(t) = \frac{1}{\sqrt{s}}\psi\left(\frac{t-\tau}{s}\right) \tag{5.16}$$

Once the wavelet function satisfies this condition, The Wavelet Transform is computed in a similar way as the STFT. The Wavelet transform of a signal $y(t)$, at the scale s and position τ, is computed by correlating y with a wavelet function ψ:

$$W(\tau,s) = \int_{-\infty}^{\infty} y(t)\psi_{\tau,s}(t)dt \tag{5.17}$$

Since wavelet transforms can measure time-frequency variations of a signal, with different time-frequency resolutions. It can also be written as frequency integration:

$$W(\tau,s) = \int_{-\infty}^{\infty} y(t)\psi^*_{\tau,s}(t)dt = \frac{1}{2\pi}\int \hat{y}(f)\hat{\psi}(t)df \tag{5.18}$$

where $\psi^*_{\tau,s}(t)$ is the complex conjugate wavelet function.

Several wavelet functions have been applied to vibration analysis, one of the most powerful is the Morlet wavelet function

$$\psi\left(\frac{t-\tau}{s}\right) = e^{\left(\frac{j2\pi f_0(t-\tau)}{s}\right)} e^{\left(\frac{-\alpha(t-\tau)^2}{s^2\beta^2}\right)} \tag{5.19}$$

where f_0, α, and β are constant.

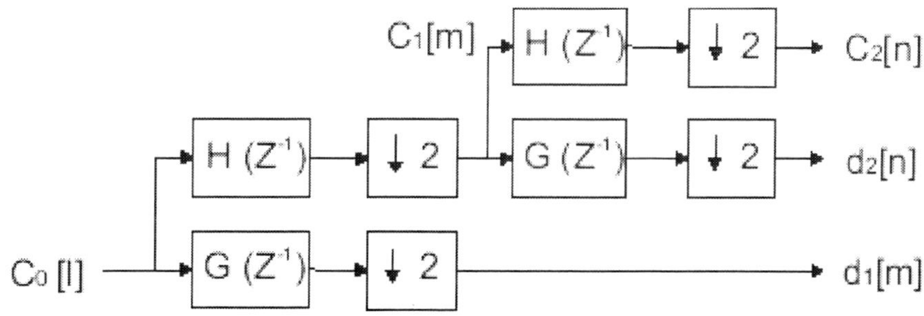

Figure 5.12. schematic procedure the wavelet calculation.

The operation is a procedure that looks for similarities between the wavelet function ψ and the vibration signal. In other words, the wavelet function is a reference, or template, and the procedure consists on the identification of which sections of the original signal matches the template. This matching is stored into a set of vectors defined as the wavelet coefficients. Figure 5.12, shows a schematic representation of the wavelet coefficient identification.

Another way of understanding this procedure is through filtering the original signal with two filters: a low-pass and a high-pass filters, namely H(z) and G(z) (Figure 5.12).

Where

$$G(Z) = (-1)^n H(1-n) \tag{5.20}$$

These two filters are constructed from the original wavelet function ψ and a scale function ϕ determined from:

$$\begin{cases} \phi(t) = \sqrt{2}\sum_n H(n)\phi(2t-n) \\ \psi(t) = \sqrt{2}\sum_n G(n)\phi(2t-n) \end{cases} \tag{5.21}$$

The coefficients of the decomposition are known as the wavelet coefficients and are calculated as:

$$\begin{aligned} c_{j,n} &= \sum_m H(2n-m)c_{m,j-1} \\ d_{j,n} &= \sum_m G(2n-m)c_{m,j-1} \end{aligned} \tag{5.22}$$

Coefficients $c_{j,n}$ represent the low frequency components of the original signal, and $d_{j,n}$ the high frequency components. The low frequency and high frequency coefficients are calculated at j level by convolving the low frequency coefficients at level $j-1$.

The signal is filtered with the low-pass filter H and high-pass filter G. The high-pass coefficients are stored at vector $d_{1,m}$. At level 2, the resulting low-pass filter signal is filtered again, and the resulting high-pass coefficients are stored at vector $d_{2,n}$. The procedure is repeated for an acceptable level of decomposition.

The wavelet function determines the ability to identify transient and nonlinear components. There is a match between the “shape” of the nonlinear and transient components with the wavelet function that is originally selected. Thus, it is fundamental to select the most appropriate wavelet function for the type of nonlinear phenomenon under study. The most commonly use wavelet functions are: Haar, Daubechies, Symlet, Coiflet and Meyer.

The basic functions are shown in next figures:

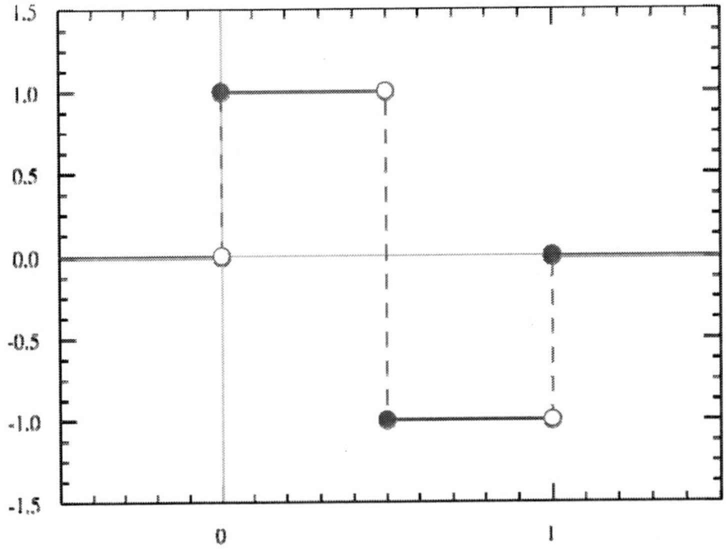

Figure 5.13. Haar Wavelet.

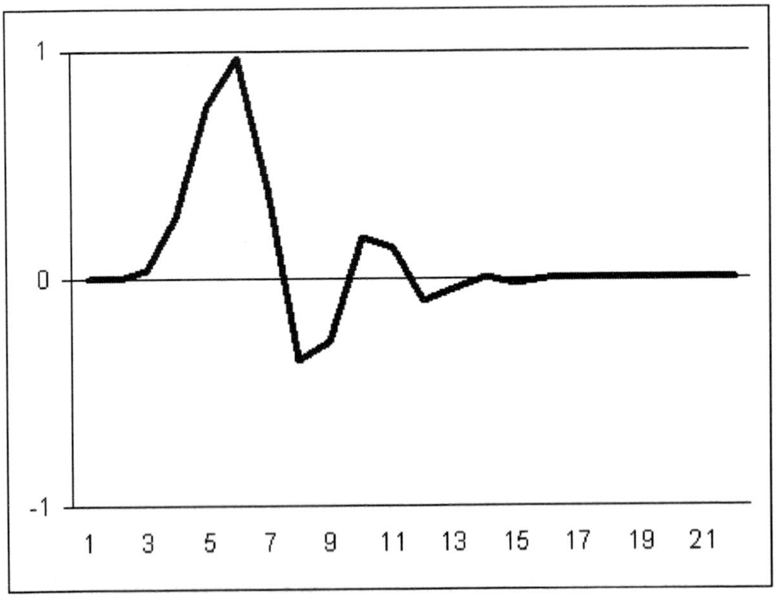

Figure 5.14. Daubechies Wavelet.

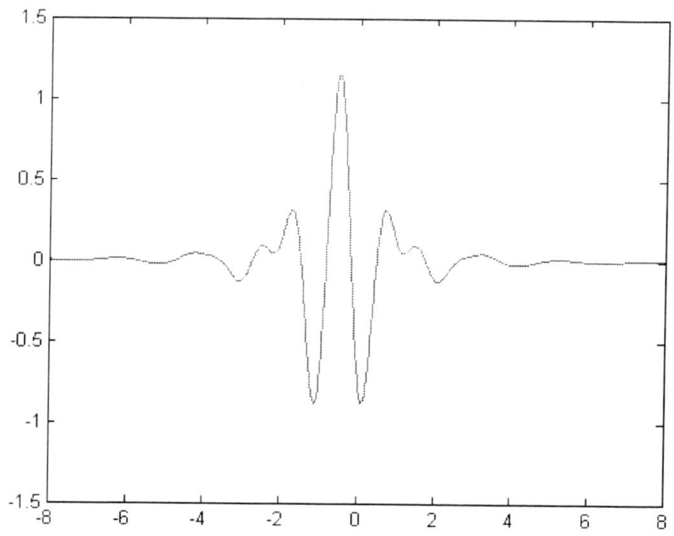

Figure 5.15. Meyer Wavelet.

The goal of the a Discrete Wavelet Transform is to decompose any arbitrary signal *y(t)* (in a discrete form it will be *y(n)*) into an infinite sum of wavelet functions with different scales *s*, and different shift time τ.

EXAMPLE

We include an example to illustrate the difference between the STFT and the wavelet transform. The output signal is simulated using eq. 5.6 with N=2. The linear terms are two sinusoidal terms, and the nonlinear coefficients are:

Linear terms

n	b_n	ω_n	ϕ_n
1	0.1	50	0
2	0.3	250	$\pi/3$

Transient terms

n	a_n	b_n	ω_n	ϕ_n
1	0.3	380	1500	0
2	0.15	300	1200	$\pi/3$

The output signal shows two transient terms, they vanished into a two steady sinusoidal waves (Figure 5.16). A traditional analysis of this signal would be a Fourier Transform, and the interpretation of the main frequencies, or "picks".

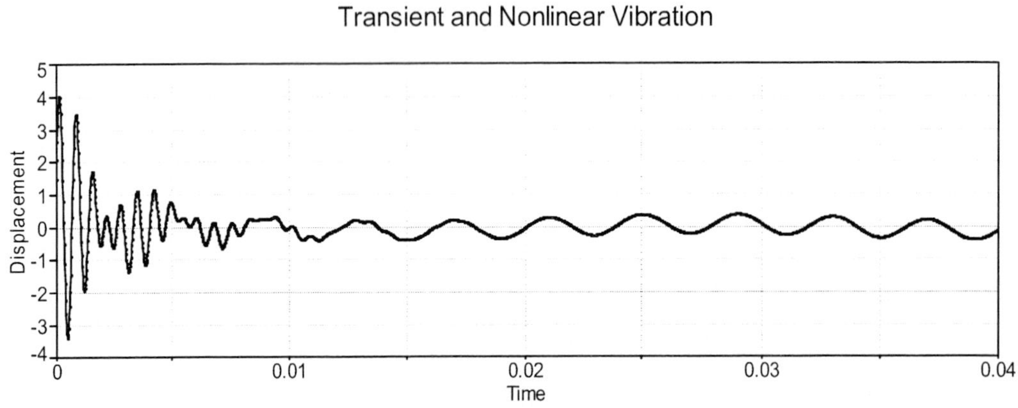

Figure 5.16. Time response.

Figure 5.17 shows the Fast Fourier Transform of the same signal, since we generated it with, we already know that there are four fundamental sinusoidal functions; two of them are damped by a transitory function, whereas the other two remain steady. The only information we get from this analysis are the four frequencies and their amplitudes. The nonlinear or transient forms are hidden under the curve.

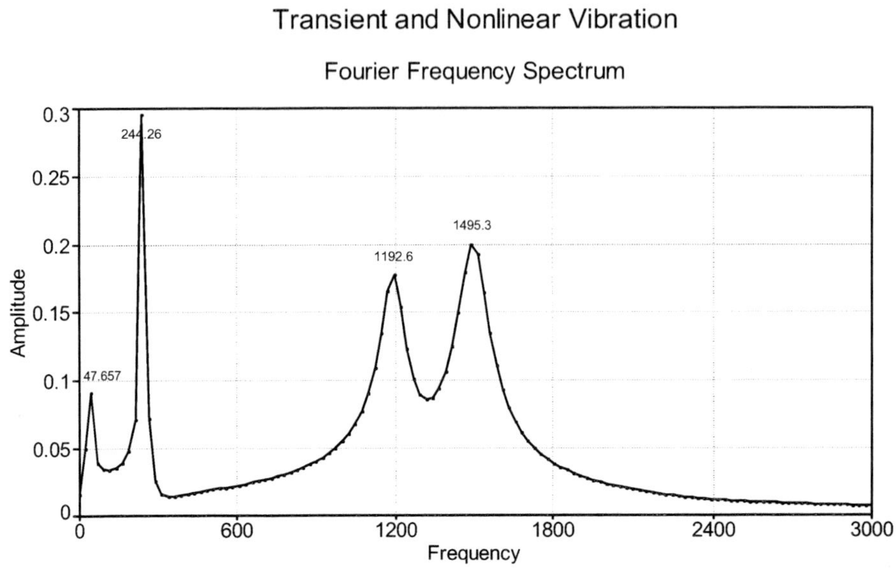

Figure 5.17. Frequency spectrum.

STFT

We first try to apply the STFT with the simplest conditions. In this case, we selected the Hann window function. As we have established before, two parameters control the Time-Frequency Map: window size and overlapping. Figure 5.18 shows a Time-Frequency Map with a small window (8 elements) and no overlapping. Under this condition, the STFT shows no significant results along the frequency dimension, and it only shows variations along the time dimension. In this sense, it is of no use the map since it gives little more information than previous analysis.

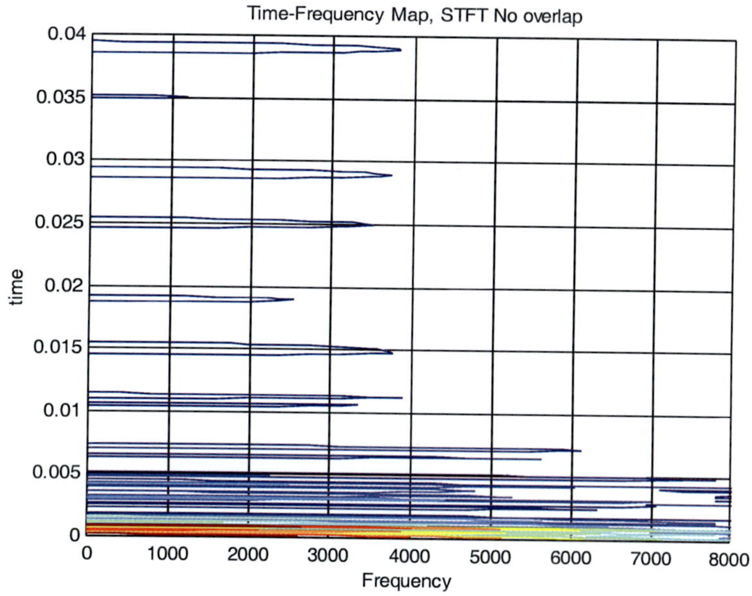

Figure 5.18. STFT Time-Frequency Map with no overlap an a window size of 8 points and a Hanning function.

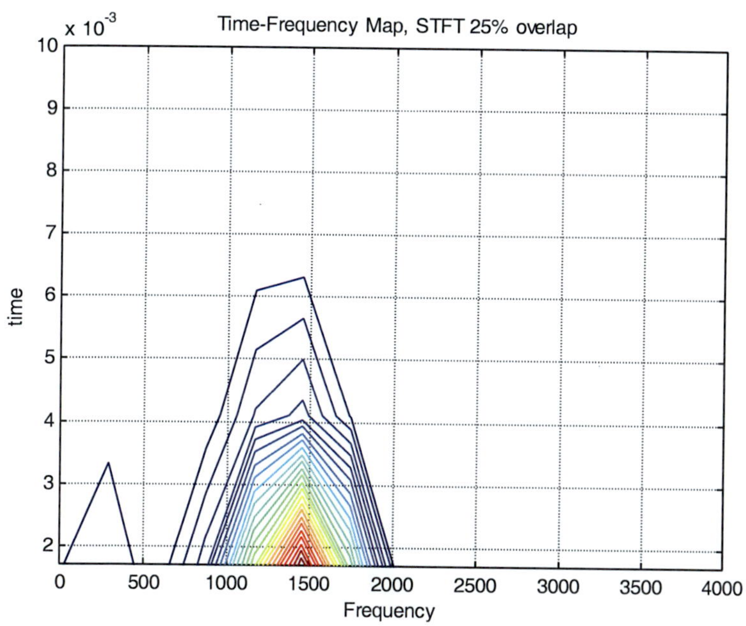

Figure 5.19. STFT Time-Frequency Map with 25% overlap an a window size of 172 points.

When we increment the window size and allow for a 25% of overlap, we obtain a map that allows us the identification of the transient effects. Since the linear frequencies and the transient frequencies are relatively close, they can not be distinguished. It also shows time variations around 250 Hz caused by the method. The advantage of this map is that the transient response is clearly identified, and it can be seen how it vanishes as time passes.

It is clear that the STFT is very sensible to the window size and the amount of overlap, but it is a fast tool that can be tuned to each particular application.

Wavelet Transform

We analyzed the same signal with two wavelets: Paul and Morlet wavelets. We selected these functions because they are similar to a vibration signal. Figure 5.20 shows the basic form of the Paul's wavelet.

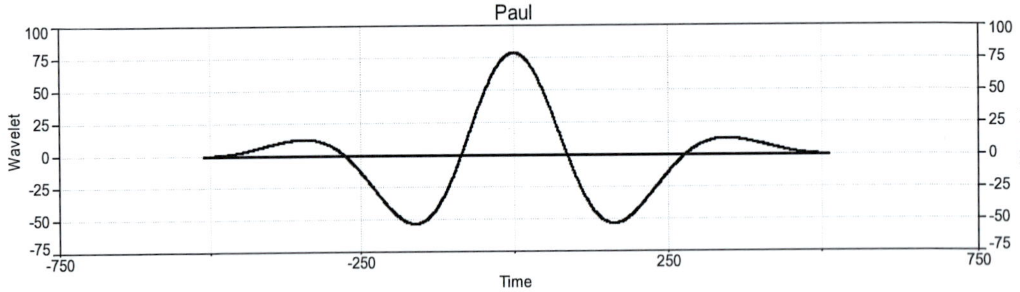

Figure 5.20. Paul Wavelet function.

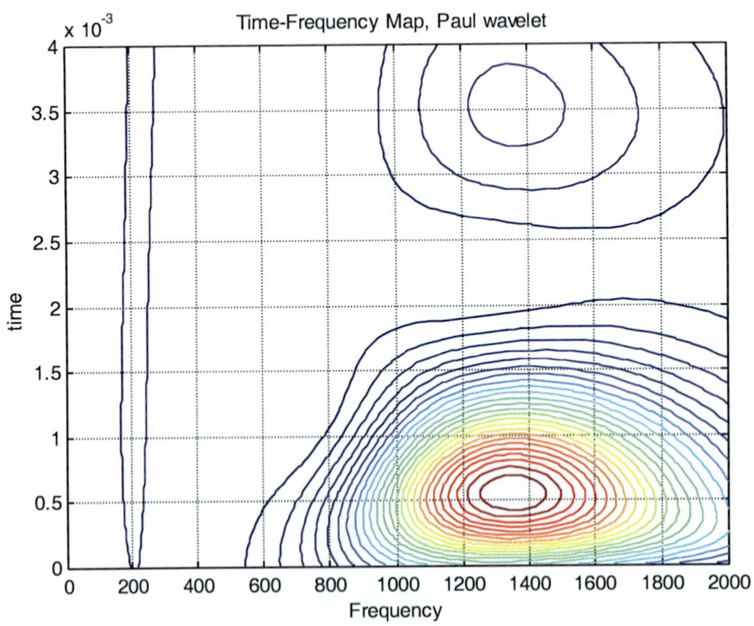

Figure 5.21. Time-Frequency Map using Paul's Wavelet function.

After applying it to the same signal, we obtained the Time-Frequency Map shown in Figure 5.21. It is similar to the last STFT map, but it has a better representation of the steady state signal component. It shows time variations around the transient signal component, but they are due to the method.

The Morlet wavelet that we applied to this signal is shown in Figure 5.22. It produces the Time-Frequency map shown in Figure 5.23.

This map provides more information than previous maps. It gives a clear identification of the two transient frequencies, and the duration time of this signal is closer to the original data. it also shows no time variations around the steady state signals, which coincide with the propose original function.

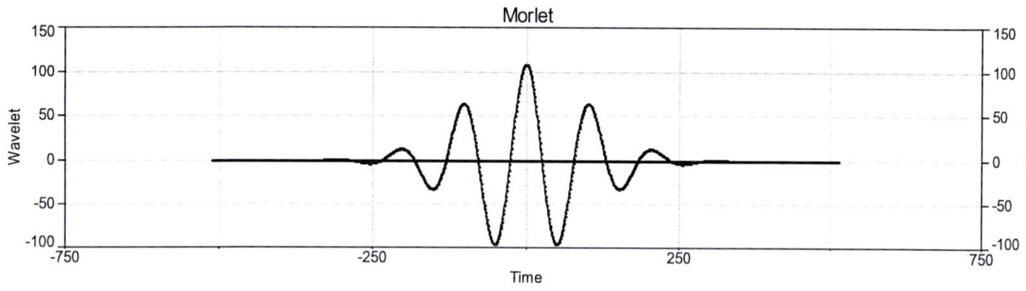

Figure 5.22. Morlet wavelet function.

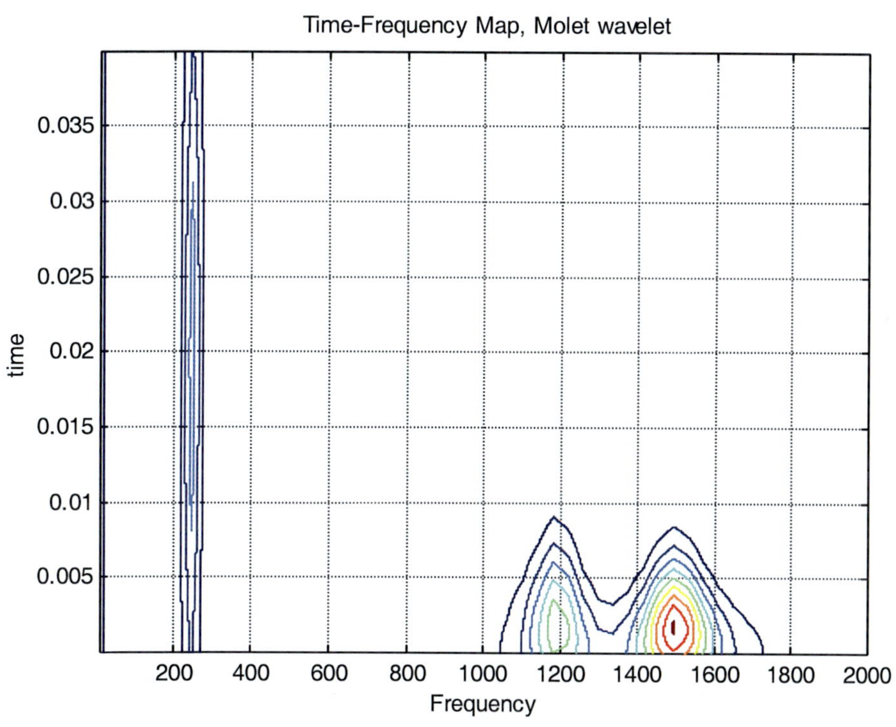

Figure 5.23. Time-Frequency Map using Morlet's Wavelet function.

We can conclude that nonlinear effects and transient signals can be identified with the phase diagram and the Time-Frequency Map. It is clear that the Morlet wavelet function is able to identify all the different information that is stored in a vibration signal.

The identification of a physical phenomenon that presents the jump effect can be obtained using the Time-Frequency Map. For this purpose, it is necessary to take measurements as the velocity of the system increments through a ramp up function and then decrements through another ramp down function. The Time-Frequency map will show amplitude variations at the critical frequency between the ramp up and the ramp down.

References

Ambarisha, V., Parker, R., 2007, "Nonlinear dynamics of planetary gears using analytical and finite element models", *Journal of Sound and Vibration,* Vol. 302, pp 577–595.

Adams, G., Tran, D., 1993. "The coefficient of restitution for a planar two-body eccectric impact" *ASME Journal of Applied Mechanics.* Vol. 60, pp. 1058-1060.

Andronov, A. et al. 1966. "*Theory of oscillators*". Dover.

Bai, C.,Zhang, H., Xu, Q., 2008, *"Effects of Axial Preload of Ball Bearing on the Nonlinear Dynamic Characteristics of a Rotor-Bearing System"*, Nonlinear Dyn, Vol 53, pp. 173–190.

Blankenship, G. W., Kahraman, A. 1996. "Gear dynamics experiments". *ASME Power Transmission and Gearing Conference.*

Berry, J., 1992 *"Use of Vibration Signature Analysis to Diagnose Machine Problems",* Technical Associates of Charlotte, Charlotte.

Castillo-Castañeda, E., Okazaki, Y. 1998. "Load variation compensation in real-time for motion accuracy of machine tools". *IEEE AMC* '98, pp. 305-309.

Chen, L., Wang, H., 2003, "Axisymmetric Vibration and Damping Analysis of Rotating Annular Plates with Constrained Damping Layer Treatments", *Journal of Sound and Vibration,* Academic Press, pp. 25-45.

Choudhury, A., Tandon, N. 1998. "A theoretical model to predict vibration response of rolling bearings to distributed defects under radial load" ASME *Journal of Vibration and Acoustics.* Vo. 120, pp. 214-220.

Cuttino et al. 1997. "Analitical and experimental identification of nonlinearities in a single-nut, preloaded ball screw". *ASME Journal of Mechanical Design.* Vol. 119, pp. 119-220.

Daidie, A., Chaib, Z., Ghosn, A., 2008, "3D Simplidied Finite Elements Analysis of Load Contact Angle in a Slewing Ball Bearing", *Journal of Mechanical Design,* Vol 130 (7), pp 1-8.

Díaz Bejarano, J., Martín Sánchez, A. 1989. "Generalized Fourier series for non-linear systems". *Journal of Sound and Vibration.* 132(2), pp. 333-341.

Ervin, R. D. 1983. *"The influence of size and weight variables on the roll stability of heavy duty trucks"*. SAE Technical Paper Series.

Fahey, S., Pratt, J. 1988. *"Frequency domain modal estimation techniques. Experimental Techniques"*. September/October 1998, pp. 33-37.

Fegelman, K., Grosh, K. 1999. "Dynamics of a vibro-impacting hinged beam: rigid and flexible models". *ASME DETC 99.* VIB-8176.

Fernandez, J., Sofonea, M., 2004, "Numerical Analysis of a Frictionless Viscoelastic Contact Problem with Normal Damped Response", *Computers and Mathematics with Applications,* Elsevier, pp. 549-568.

Filipich, C. P., Rosales, M. B. 1996. A brief discussion on some energy principle. *Journal of Vibration and Control.* Vol. 2, pp. 275-287.

Fox, B. et al. 2001. "Numerical computation of differential equations for non-linear dynamics of multibody systems involving contact forces". *ASME Journal of Mechanical Design.* Vol. 123, pp. 272-280.

Fung, E., Yau, D., 2003, "Vibration Characteristics of a Rotating Flexible Arm with ACLD Treatment", *Journal of Sound and Vibration,* Academic Press, pp. 165-182.

Gao, R., Yan, R., 2006 "Non Stationary signal processing for bearing health monitoring", International Journal of Manufacturing Research, vol 1., No. 1.

Ghafari, S., Golnaraghi, F., Ismail, F., 2008, *"Effect of Localized Faults on Chaotic Vibration of Rolling Element Bearings*", Nonlinear Dyn, Vol 53, pp. 287–301.

Green, R. 1996. *"Machinery's handbook".* New York Industrial Press.

Grigorescu, R., Gafitanu, M., 2000, "Constructive Optimization Criteria to Increase Bearing Service Life", *Tribology Letters,* Vol 8, pp. 57–62.

Guckenheimer, J., Holmes, P. 1983. *"Nonlinear oscillations, dynamical systems and bifurcation of vector fields".* Springer-Verlag.

Gupta, T., Gupta, K., Sehgal, K., 2008, "Nonlinear Vibration Analysis of an Unbalanced Flexible Rotor Supported By Ball Bearings with Radial Internal Clearance", *Proceedings of GT2008 ASME Turbo Expo 2008,* Berlin, Germany.

Halse, C., Wilson, R., di Bernardo, M., Homer, M., 2007, "Coexisting solutions and bifurcations in mechanical oscillators with backlash", *Journal of Sound and Vibration,* Vol. 305, (4-5), pp. 854-885.

Harris, T., 1991, *Rolling Bearing Analysis, Wiley-Interscience,* New York.

Harsha, S., Sandeep, K., Prakash, R., 2004, "Nonlinear Dynamic Response of a Rotor Bearing System Due to Surface Waviness", *Nonlinear Dynamics*, Vol 37, pp 91–114.

Hilborn, R. C. 1994. *"Chaos and nonlinear dynamics"* Oxford University Press.

Hilton, H., Yi, S., 1999, "Generalized Viscoelastic 1-DOF Deterministic Nonlinear Oscillators", *Journal of Non-Linear Mechanics,* Vol. 1-22.

Iyer, N., Qiu, H., Yan, W., 2007, "Early Detection of Lubrication Anomalies in Oil-Lubricated Bearings", *Proceedings of GT2007 ASME Turbo Expo 2007,* Montreal, Canada.

Jerrelind, J., Stensson, A., 2000, "Nonlinear dynamics of parts in engineering systems", *Chaos, Solitons and Fractals* Vol. 11, pp 2413-2428.

Johnson, W. 1982. *"Slip line field for metal-deformation processes".* Pergamon Press.

Khan A., Vyas, N. 2001, "Application of Volterra and Wiener Theories for Nonlinear Parameter Estimation in a Rotor-Bearing System", *Nonlinear Dynamics* Vol (24), pp. 285–304.

King, M. E. 1997. *"Nonlinear modes in systems with strong nonlinearities".* ASME DETC 97. VIB-3949.

Liao, N., Lin, J., 2001. "A new method for the analysis of deformation and load in a ball bearing with variable contact angle" *ASME Journal of Mechanical Design.* Vol. 123, pp. 304-312.

Lieh, J. "A multibody dynamic program for truck simulation". *Heavy vehicle dynamics and simulation in braking, steering and suspension systems.* SAE.

Lozano, A., Lopez, C., Jauregui, J., Fawcett, J., 1991, "Transverse Vibration of a Flexible Strip Travelling Between Two Free Supports", *Proc. of the 8th World Congress on the Theory of Machines and Mechanisms,* Prague, Czechoslovakia, pp. 1107-1110.

Lozano, A., Lopez, C., Jauregui, J., 1995, "Viscoelastic Beam Vibration Under Axial Tension Variations", *Proc. of the 9th World Congress on the Theory of Machines and Mechanisms*, Milan, Italy, pp. 1136-1139.

Luo, A. Chen, L., 2005 "Periodic motions and grazing in a harmonically forced, piecewise, linear oscillator with impacts " Chaos, Solitons and Fractals Vol 24, pp 567–578.

Marion, M., Temam, R. 1988. "Nonlinear Galerkin methods". *SIAM Journal of numerical analysis".* Vol. 26, No. 5, pp. 1139-1157.

Mason, J., Homer, M., Wilson, R., 2007, "Mathematical models of gear rattle in Roots blower vacuum pumps", *Journal of Sound and Vibration,* Vol. 308, (3-5), pp. 431-440.

Mesquita, A., Coda, H., 2003, "New Methodology for the Treatment of two Dimensional Viscoelastic Coupling Problems", *Computer Methods in Applied Mechanics and Engineering,* Elsevier, pp.1911-1927.

Moon, F. C., Holmes, P. J. 1979. "A magnetoelastic strange attractor". *Journal of Sound and Vibration.* 65(2), pp. 275-296.

Nayfeh, A. H. et al. 1995. "Nonlinear normal modes of a cantilever beam". *ASME Journal of Vibration and Acoustics.* Vol. 117, pp. 477-481.

Nayfeh, A., Mook, D. 1979. *"Nonlinear oscillations*". John Wiley & Sons.

Nayfeh, A. H., 1981. *"Introduction to perturbation techniques".* John Wiley & Sons.

Newland, D., 1993, *An introduction to Random Vibrations, Spectral and Wavelet Analysis,* Prentice Hall, London.

Parey, A., El Badaoui, M., Guillet, F., Tandon, N., 2006, "Dynamic modeling of spur gear pair and application of empirical mode decomposition-based statistical analysis for early detection of localized tooth defect", *Journal of Sound and Vibration,* Vol. 294, (3), pp 547-561.

Pesheck, E., Pierre, C. 1997. "A global methodology for the modal reduction of large nonlinear systems containing quadratic and cubic nonlinearities". *ASME DETC 97*. VIB-3952.

Press et al. 1992. *"Numerical recipes"* Cambridge University Press.

Raju, P., Richa, M., 2003, *"Impact and Dynamic Response of High-Density Structural Foams Used as Filler inside Circular Steel Tube",* Composite Structures, Elsevier, pp. 291-302.

Rand, R., Armbruster, D. 1987. *"Perturbation methods, bifurcation theory and computer algebra"* Springer –Verlag.

Rubio, E., Jauregui, J., Gonzalez, O., 2006 "Aplicación de la Transformada Wavelet en la Detección de Defectos Causados por Vibrado en Piezas Cilíndricas Rectificadas", *Ingeniería Mecánica Tecnología y Desarrollo,* Vol 2(2), pp. 47-52.

Sawicki, J., Johansson, S., Rumbarger, J., Sharpless, R., 2007, "Fatigue Life Prediction for Large-Diameter Elastically-Constrained Ball Bearings", *Proceedings of GT2007 ASME Turbo Expo 2007,* Montreal, Canada.

Setto, W. W. 1964. "Mechanical vibrations" Shaum's Series. McGraw-Hill.

Shabana, A. A. 1998. *"Dynamics of multibody systems".* Cambrigde University Press.

Smith, D. 1983. *"Gears and their vibration".* The Macmillan Press.

Smith, M. H. et al. 1995 "Adaptive control strategies for a precision machine tool axis". *Precision Engineering* 17, pp. 192-206.

Stoker, J. J. 1950. *"Nonlinear vibrations in mechanical and electrical systems"*. John Wiley & Sons.

Strogatz, S. H. 1994. *"Nonlinear dynamics and chaos"*. Perseus Books.

Theodossiades, S., Tangasawi, O., Rahnejat, H. 2007, "Gear teeth impacts in hydrodynamic conjunctions promoting idle gear rattle", *Journal of Sound and Vibration,* Vol. 303, (3-5), pp. 632-658.

Timings, R.L. 1995, "Tecnología de la fabricación". Alfaomega grupo editor.

Vakakis, A. F. et al. 1995. *"Normal modes and localization in nonlinear systems"* Wiley Interscience.

Vela, L., Jauregui, J., Gonzalez, O., Herrera, G., Lozano, A., 2008, *"Modeling of Machining Proceses for Predictive Analysis of Self-exited Vibrations"*, Ingeniería Mecánica Tecnología y Desarrollo, Vol 3(1), pp. 1-9.

Velex, P. et al. 1996. *"Some numerical methods for the simulation of geared transmission dynamic behavior formulation and assessment"*. ASME Power Transmission and Gearing Conference.

Young, W. C. 1989. *"Roark's formulas for stress and strain"*. McGraw-Hill.

Yu, J., Bentley, D., Goldman, P., Dayton, K., Van, S., 2002, "Rolling Element Bearing Defect Detection and Diagnostics Using Displacement Transducers", *Journal of Engineering for Gas Turbines and Power* (ASME), Vol. 124, pp 517-527.

Watson, M., Sheldon, J., Amin, S., Lee., H., Byington, C., Begin., M., 2007, "A Comprehensive High Frequency Vibration Monitoring System for Incipient Fault Detection of Gears, Bearings and Shaft/Couplings in Turbine Engines and Accessories", *Proceedings of GT2007 ASME Turbo Expo 2007,* Montreal, Canada.

Zhang, D., Xie, Y., Feng, Z., 2008, "An Investigation on Dynamic Characteristics of a High Speed Rotor With Complex Structure for Microturbine Test Rig", *Proceedings of GT2008 ASME Turbo Expo* 2008, Berlin, Germany.

INDEX

A

accuracy, 36, 51, 93, 133
aid, viii, 5, 33, 94
algorithm, 111
alternative, viii, 22, 122
alternatives, 36
amplitude, 25, 28, 39, 59, 60, 61, 72, 88, 90, 96, 132
analog, 88, 90, 114
angular velocity, 40, 44, 46
appendix, 18
application, ix, 1, 15, 20, 33, 45, 48, 56, 58, 59, 63, 71, 76, 93, 106, 111, 113, 114, 122, 123, 130, 135
assessment, 136
ASTM, 103
asymptotically, 6
attractors, 6
automobiles, 93

B

backlash, 33, 34, 36, 37, 51, 56, 134
bandwidth, 124
beams, 63, 80, 88, 90, 93, 112
behavior, vii, viii, 1, 4, 25, 28, 34, 40, 41, 42, 47, 48, 58, 63, 71, 77, 78, 79, 80, 87, 88, 90, 93, 95, 97, 111, 112, 113, 136
bending, 37, 67, 68, 88
bifurcation, 1, 2, 10, 11, 12, 13, 14, 15, 16, 54, 56, 134, 135
bifurcation theory, 135
blocks, 121

C

Canada, 134, 135, 136
chaos, 1, 56, 136
chaotic behavior, viii
circulation, 51
classical, 43
classification, 11
compensation, 133
components, 56, 60, 69, 120, 125, 126
computation, 134
concentration, 56
conditioning, 23
configuration, 1, 10, 118
Congress, 135
construction, 29, 51, 93, 95, 111, 121
control, 12, 111, 113, 128, 136
Coulomb, vii, 57
cutting force, 55
cycles, 56, 88

D

damping, vii, viii, 28, 35, 40, 51, 57, 67, 80, 87, 88, 96, 97, 98, 104, 106, 107, 108, 109
data processing, 123
decay, 120
decomposition, 121, 125, 126, 135
defects, 36, 37, 44, 45, 48, 133
definition, 29, 39, 45, 59, 67, 108, 111, 112, 121
deformation, 34, 43, 44, 51, 52, 63, 65, 69, 80, 88, 89, 90, 97, 98, 99, 100, 134
degrees of freedom, 36, 52, 64, 84, 94, 95, 100, 101
detection, 135
deviation, 20, 34
differential equations, viii, 1, 15, 25, 36, 111, 134
dimension number, 10
discontinuity, vii, 33, 37, 42, 48, 56
discretization, 122

displacement, vii, 8, 29, 42, 44, 48, 51, 52, 53, 58, 59, 64, 67, 68, 69, 71, 73, 77, 85, 87, 88, 89, 90, 94, 99, 100, 102, 113
distortions, 87
distribution, 29, 31, 43, 120
duration, 131
dynamical systems, vii, 134

E

elastic deformation, 34, 51, 97
elasticity, 63
elastomers, 93
electrical system, 136
empirical mode decomposition, 135
energy, 48, 56, 60, 65, 66, 69, 99, 100, 107, 134
engines, 59, 93
equilibrium, 1, 6
evolution, 1
excitation, vii, viii, 25, 28, 36, 45, 47, 55, 71, 74, 75, 76, 77, 90, 96, 112, 113

F

fabrication, 36
fabrication errors, 36
failure, 56
family, 4
fault detection, 111
faults, 48, 49
FFT, 33, 58, 60, 111, 123
film, 35, 36, 40
film thickness, 40
filters, 125
finite differences, 95
floating, 95
flow, 29
fluctuations, 36, 56
fluid, 42, 95
Fourier, 17, 24, 31, 33, 39, 44, 45, 52, 111, 119, 120, 121, 127, 128, 133
Fourier analysis, 33
freedom, 36, 52, 63, 64, 71, 80, 84, 87, 94, 95, 100, 101, 102
frequency resolution, 121, 122, 124
friction, vii, 40, 43, 51, 57, 93
fulfillment, 63

G

gas, 42, 63
gas turbine, 63
Gaussian, 122
Germany, 134, 136
graph, 29, 61
grazing, 135
grids, 122
growth, 29

H

harmonic frequencies, 58
harmonics, 58, 60
health, 134
heat, 34, 107
height, 38
high temperature, 97
Hilbert, 1, 60
Hilbert space, 1
homogeneity, 113
housing, 36, 43
hub, 107
hydrodynamic, 136

I

identification, vii, ix, 36, 43, 48, 58, 111, 120, 124, 125, 130, 131, 132, 133
identity, 17, 31
indication, 40, 47, 54, 59, 88, 114
indicators, 113
industrial, 33
industrial application, 33
inertia, 107
infinite, 15, 127
instabilities, 29, 48, 50
integration, 106, 114, 124
interference, 120
intrinsic, 42
isolation, 94
Italy, 135
iteration, 122

K

kinematics, 43, 95
kinetic energy, 65, 66, King, 134

L

laws, 1
limitation, 122
limitations, 60, 87, 94, 96, 122

linear, vii, viii, 3, 4, 5, 6, 9, 16, 22, 28, 33, 40, 44, 48, 58, 63, 64, 65, 67, 68, 69, 71, 72, 74, 77, 79, 80, 82, 84, 87, 95, 97, 100, 105, 111, 112, 114, 118, 119, 120, 127, 130, 133, 135
linear model, 58
linear systems, viii, 133
localization, 136
London, 135
long period, 40
lubrication, 34

M

M1, 36
machinery, viii, 56, 60, 93, 111
machines, viii, 34, 56, 59, 93, 111
magnetic, 42
magnetoelastic, 135
maintenance, 33
management, 104
manufacturing, 37, 39, 41, 93, 111
matrix, 5, 6, 7, 8, 9, 69, 95, 96, 100, 102, 105, 106, 109, 114
measurement, vii, viii, 60, 104
mechanical behavior, 97
mechanical energy, 107
metals, viii, 97
microprocessors, 111
modeling, vii, viii, 135
models, viii, 36, 57, 67, 88, 93, 111, 133, 135
modulation, 118, 119
modulus, 104
Moon, 1
Morlet wavelets, 130
motion, 33, 34, 35, 36, 56, 57, 64, 67, 68, 70, 71, 98, 100, 101, 102, 133
movement, 56, 57, 93
multidimensional, 95

N

natural, 71, 75, 77, 79, 96, 107, 108
New York, 134
Newton, 1, 4, 95
Newtonian, viii, 95
nodes, 6, 64, 67, 68, 70, 94, 101
noise, 34, 36, 93, 118, 119
nonlinear dynamic systems, 55
non-linear dynamics, 134
nonlinear systems, vii, viii, ix, 29, 33, 111, 113, 114, 135, 136
nonlinearities, vii, viii, 42, 44, 67, 114, 133, 134, 135
non-linearity, 15
normal, 11, 57, 58, 135
numerical analysis, 135

O

oil, 35, 36, 40
one dimension, 120
orbit, 29
orientation, 29
oscillations, 134, 135
oscillator, 2, 4, 6, 7, 9, 17, 18, 19, 20, 21, 22, 24, 27, 28, 135

P

parameter, 10, 12, 33, 36, 42, 108, 120, 124
partial differential equations, 95
partition, 101
passenger, 93
passive, viii, ix, 93, 94, 96, 98, 105
periodic, 7, 10, 29, 34, 39, 40, 45, 58, 90, 119, 120
perturbation, 5, 10, 15, 23, 135
perturbation theory, 15
phase diagram, ix, 3, 4, 6, 7, 8, 10, 33, 40, 47, 48, 49, 54, 55, 59, 71, 72, 76, 77, 84, 87, 90, 111, 112, 114, 115, 116, 117, 118, 132
phase space, 120
physics, 1
pitch, 34, 35, 36, 39, 40, 43, 46
planar, 133
planetary, 133
plastic, vii, 63
plastic deformation, 63
Poincaré, 1, 29, 30, 31
polynomial, viii, 33, 58, 64, 68, 94, 100, 102, 105, 120
polynomial functions, viii
potential energy, 69
power, 34, 71, 107
prediction, 36
pressure, 40, 65, 100
probability, 120
probability distribution, 120
probe, 103
program, 135
propagation, 5
property, 34, 113, 114, 122
protection, 94
prototype, 12
pulses, 122
pumps, 34, 135

Q

quantum, 120
quantum mechanics, 120

R

race, 45, 46
radius, 3, 39, 40, 63, 80
random, viii, 119
range, viii, 10, 28, 34, 51, 55, 96, 108
real time, 18, 48, 58
reality, 34
reciprocity, 114
redundancy, 122
regular, 15
rejection, 6
relationship, 33, 97, 114
relativity, 20
resistance, 56
resolution, 121, 122
restitution, 133
rods, 107
rolling, vii, 34, 40, 42, 43, 44, 45, 46, 47, 48, 51, 52, 133
rotational inertia, 36
rubber, viii, 67, 97, 103, 104, 107, 108, 109

S

SAE, 133, 135
sample, 121
sampling, 122
segmentation, 120
Self, 136
sensitivity, 1
sensors, 111
separation, 19
services, iv
shape, 3, 10, 29, 34, 47, 48, 50, 58, 71, 77, 84, 87, 88, 90, 112, 113, 114, 116, 126
shear, 95, 98
shock, 97
shocks, 93
short period, 57
sign, 8, 29, 114
signals, 41, 56, 60, 111, 116, 118, 120, 131, 132
simulation, 135, 136
singular, 26
spatial, 64
spectral analysis, ix, 120
spectrum, 33, 34, 58, 113, 120, 121, 122, 128
speed, 28, 34, 36, 41, 48, 49, 58, 59, 60, 61, 63, 111
springs, viii, 43, 53, 54, 94
stability, 4, 5, 10, 12, 31, 87, 111, 133
stages, 60
standard deviation, 104
statistical analysis, 135
steady state, 26, 120, 131
steel, 88, 93
stiffness, vii, viii, ix, 8, 29, 33, 34, 36, 37, 38, 39, 40, 42, 43, 44, 45, 47, 48, 51, 52, 53, 56, 58, 63, 67, 71, 90, 94, 95, 96, 101, 104, 105, 106, 108, 114, 118
strain, 63, 64, 67, 68, 95, 97, 100, 102, 105, 136
strange attractor, 135
strategies, 136
stress, 37, 58, 95, 97, 105, 107, 136
structural defects, 44
Sun, 1
superposition, 112, 113, 120
supply, 107

T

teeth, 34, 35, 36, 37, 38, 39, 40, 41, 136
temperature, vii
threshold, 96, 108
time resolution, 121
torque, 36, 39, 108
trajectory, 1, 3, 6, 29, 34
transducer, 48
transfer, 113
transformation, 101, 120, 121
translation, 43, 44, 45, 48, 51
transmission, 40, 42, 136
transmits, 34, 36
trucks, 133
trust, 63, 64, 65, 66, 67, 93, 102

U

uncertainty, 122
uniform, 80
universe, 1
uti, 66, 99

V

vacuum, 135
values, 12, 28, 59, 61, 96, 104, 108, 109, 124
variables, viii, 29, 133
variation, 35, 36, 43, 44, 52, 58, 60, 113, 133

vector, 1, 3, 10, 69, 100, 101, 120, 121, 122, 125, 134
vehicles, 93
velocity, vii, 29, 35, 40, 44, 46, 59, 69, 71, 73, 77, 85, 87, 88, 89, 90, 99, 121, 132
vibration, viii, ix, 33, 34, 36, 40, 41, 44, 47, 48, 56, 58, 60, 71, 77, 78, 79, 82, 83, 84, 85, 86, 87, 88, 90, 91, 92, 93, 94, 95, 96, 98, 102, 103, 107, 122, 123, 124, 125, 130, 132, 133, 135
viscoelastic properties, 101, 103

W

wavelet, 60, 122, 124, 125, 126, 127, 130, 131, 132
wavelet analysis, 60
Wavelet Transform, 123, 124, 127, 130
wavelets, 58, 113, 123, 124, 130
wind, 63
windows, 122